The Leadership Algorithm

What It Takes to Lead in the Age of Data, AI, and Intelligent Systems

Shrey Shivam

Apress®

The Leadership Algorithm: What It Takes to Lead in the Age of Data, AI, and Intelligent Systems

Shrey Shivam
Dublin, Ireland

ISBN-13 (pbk): 979-8-8688-2896-6 ISBN-13 (electronic): 979-8-8688-2897-3
https://doi.org/10.1007/979-8-8688-2897-3

Managing Director, Apress Media LLC: Welmoed Spahr
Acquisitions Editor: Shivangi Ramachandran
Editorial Assistant: Jessica Vakili
Copy Editor: Kezia Endsley

Cover designed by eStudioCalamar

Cover image designed by kjpargeter on Freepik

Distributed to the book trade worldwide by Springer Science+Business Media New York, 1 New York Plaza, New York, NY 10004. Phone 1-800-SPRINGER, fax (201) 348-4505, e-mail orders-ny@springer-sbm.com, or visit www.springeronline.com. Apress Media, LLC is a Delaware LLC and the sole member (owner) is Springer Science + Business Media Finance Inc (SSBM Finance Inc). SSBM Finance Inc is a **Delaware** corporation.

For information on translations, please e-mail booktranslations@springernature.com; for reprint, paperback, or audio rights, please e-mail bookpermissions@springernature.com.

Apress titles may be purchased in bulk for academic, corporate, or promotional use. eBook versions and licenses are also available for most titles. For more information, reference our Print and eBook Bulk Sales web page at http://www.apress.com/bulk-sales.

Any source code or other supplementary material referenced by the author in this book is available to readers on GitHub. For more detailed information, please visit https://www.apress.com/gp/services/source-code.

If disposing of this product, please recycle the paper

For my wife—your strength through adversity inspired this journey more than words can express.

To our families, whose unwavering support, courage, and resilience helped us navigate the challenges that shaped this journey.

Table of Contents

About the Author

Shrey Shivam is a visionary data and AI leader with over a decade of experience driving large-scale data transformation across global enterprises. As Senior Director and Head Of Data, Analytics, and AI for EMEA at Fiserv, he specializes in building data products and AI-driven platforms that accelerate customer growth, operational excellence, and business innovation.

He has a proven track record of defining growth strategies, scaling high-performing cross-functional product and engineering teams, and commercializing innovative data solutions across industries, including fintech, insurance, retail, ecommerce, digital media, investment banking, as well as transaction-led marketplaces spanning music, food, lifestyle, news, legal, and travel. His expertise spans product management, data strategy, machine learning and AI, and modern data platforms, with deep hands-on experience across cloud, big data, and AI ecosystems.

Prior to Fiserv, Shrey held leadership roles at Prudential Financial and Times Internet and co-founded a technology startup, bringing together entrepreneurial thinking with deep technical acumen. His diverse experience across startups, consulting, and enterprise environments gives him a unique perspective on building scalable, customer-centric solutions.

Shrey holds a BTech in Information Technology from Cochin University of Science and Technology and an MBA from Queen's University Belfast. He is also the co-author of the book *Building an Enterprise Chatbot* by Apress and brings a distinctive perspective on aligning emerging technologies with real-world business impact in the age of AI.

About the Technical Reviewer

Dipesh Singh leads the agent platform at JP Morgan Chase, where his team builds AI services and products for document understanding, RAG, NL2SQL, and intelligent document workflows that have exceeded public benchmarks in extraction and comprehension.

His work is shaped by a simple conviction: AI that matters is AI that ships, stays accurate, and gets better with use. He has built and productionized systems where humans actively shape model improvement through structured feedback, rigorous evaluation, and continuous iteration, rather than just providing oversight. The result is AI that earns trust over time, not just at launch.

Dipesh thinks in platforms and obsesses over pushing the state-of-the-art into production, closing the gap between what's possible in research and what actually works at enterprise scale.

When he's not building AI systems, he's at home in New Jersey being outwitted by his three-year-old daughter who, he suspects, already has a better feedback loop than most models he's deployed.

Acknowledgments

This book has been shaped by years of working with colleagues, leaders, and teams across data, analytics, and AI-driven transformation. I am grateful to those I have worked alongside in different organizations, whose perspectives, challenges, and debates have deeply influenced my thinking about what it truly takes to turn data and AI into measurable business value. Much of what is shared here is grounded in real-world experience - delivering large-scale data initiatives, navigating organizational complexity, and learning that transformation is as much about leadership, alignment, and culture as it is about technology.

I am also thankful to the broader data and AI community, whose collective work continues to advance this field at an extraordinary pace. The ideas in this book have evolved through years of shared learning across practitioners, researchers, and executives.

Any value this book creates is a reflection of the many people who have contributed, directly and indirectly, to the thinking behind it.

Introduction

We are entering one of the most significant shifts in business history.

For decades, software helped organizations automate processes, store information, and improve efficiency. Today, a new generation of AI is changing the equation entirely. Intelligent agents, powered by advances in generative and agentic AI, are beginning to reason, act, make decisions, and execute workflows with increasing autonomy. These systems are no longer limited to answering questions or generating content; they are evolving into digital collaborators capable of transforming how organizations operate.

Across boardrooms around the world, leaders are asking urgent questions:

- How do we use AI to create a competitive advantage?
- How do we prepare for a future where software can act independently?
- How do we avoid being disrupted by companies that move faster than we do?

But amid the excitement surrounding AI, many organizations are confronting an uncomfortable reality.

Despite billions invested in cloud platforms, analytics tools, machine learning, and AI initiatives, most companies still struggle to generate meaningful business value from data. AI pilots fail to scale. Dashboards proliferate while decision-making remains slow. Data remains fragmented across systems and teams. Governance is seen as a bottleneck instead of an enabler. And many executives continue to chase AI transformation without fixing the foundational problems that undermine it.

This is the paradox at the heart of modern data leadership.

The future promises intelligent organizations powered by autonomous AI agents, yet many businesses still lack the data foundations, operating models, and leadership alignment necessary to support that future.

That is why this book exists.

The Leadership Algorithm is a practical and strategic guide for Chief Data Officers (CDOs), Chief Data and AI Officers (CDAIOs), analytics leaders, technology executives, product leaders, and transformation professionals navigating the rapidly evolving world of data and AI.

At its core, this book argues that success with AI is not primarily a technology problem, it is a leadership, organizational, and data strategy problem.

The organizations that succeed in the age of agentic AI will not necessarily be the ones with the most advanced models or the largest technology budgets. They will be the ones that build strong data foundations, align business and technology strategies, create cultures of trust and accountability, and design organizations capable of adapting continuously.

This book explores how leaders can make that transition.

Rather than focusing purely on technical architectures or theoretical frameworks, *The Leadership Algorithm* examines the real-world challenges that data leaders face every day: organizational silos, fragmented ownership, poor governance, resistance to change, unrealistic executive expectations, unclear ROI, and the growing pressure to operationalize AI at scale.

Most importantly, this book is designed to help leaders think differently.

The future will belong to organizations that learn faster, adapt quicker, and align data, technology, and leadership more effectively than their competitors.

This book is about how to build those organizations.

CHAPTER 1

The Data Illusion: Why Most Companies Get It Wrong

Richard Marshall leaned back in his chair, watching the city lights flicker through the glass walls of the 32nd floor conference room. It was 7:00 PM, another late evening. Across from him sat the company's senior leadership team: Lisa from Strategy, Tom from Marketing, and Priya from Finance. The quarterly business review had stretched longer than expected, but Richard wasn't surprised. Every conversation about their data and AI initiatives felt like running on a treadmill—plenty of motion, but no real progress.

"We've spent millions upgrading our data platform," Richard began, his voice calm but edged with frustration. "Yet, here we are—sales projections are still off, customer churn remains a mystery, and I keep hearing that teams can't find the data they need."

Lisa, the Chief Strategy Officer, sighed and folded her arms. "It's not the technology, Richard. We have dashboards for everything—market trends, product performance, and customer segments. But the insights aren't shaping decisions. We're drowning in data but starving for clarity."

S. Shivam, *The Leadership Algorithm*, https://doi.org/10.1007/979-8-8688-2897-3_1

Tom, the Marketing Director, jumped in. "Exactly. We launched a customer personalization project last year. The algorithms were supposed to improve our campaign targeting. Instead, we got hundreds of charts telling us that 'engagement is up by 12%.' What does that even mean for revenue? We still can't connect the dots between marketing spend and actual business outcomes." He paused, then added, more thoughtfully, "And the frustrating part is, it feels like every new use case starts from scratch. New data pulls, new models, new interpretations. If we're investing this much, shouldn't it get easier over time? Shouldn't the next initiative cost less and deliver faster? Right now, there's no sense of compounding value, just repeated effort."

"Exactly," Priya, CFO, nodded. "If the foundation were right, each new use case should reduce marginal cost—not increase it. That's the return I'm not seeing." She adjusted her glasses, her analytical mind already calculating the growing cost of these "data-driven" efforts. "And let's not forget how much we're spending. Every month, there's another proposal for a new tool or an AI model. But where's the ROI? We're collecting more data than ever, but it's not making us smarter. I can't approve more funding without a clear business impact."

Priya paused, then added more deliberately, "And it's not just about the cost anymore. We operate in a highly regulated industry with compliance, governance, and regulatory oversight; they're all starting to push back. They're asking how our models make decisions, where the data is coming from, and whether we can explain it consistently across the organization. The problem is, we can't. Each function has its own definitions, its own logic, and its own version of the truth. Even our governance and compliance teams are struggling to reconcile the data across systems. If we scale this without fixing that, we're not just wasting money—we're exposing ourselves to serious regulatory risk."

Richard pinched the bridge of his nose. This was the illusion of data—an expensive mirage. The company had embraced cutting-edge technology, yet the promised transformation remained elusive. Despite the glossy dashboards and advanced models, the core business questions were still unanswered.

"What if we've been looking at this the wrong way?" Richard asked quietly.

The room fell silent.

"We keep throwing technology at the problem—more tools, more algorithms—but we're missing the point. Data isn't the solution. It's how we align that data with business decisions. If we don't change how we structure our teams and define what we're solving for, we'll keep spinning our wheels." Lisa leaned forward, curious. "So, you're saying the issue isn't about having more data—it's about how we use it?"

"Exactly," Richard nodded.

"Right now, data lives in silos. Strategy, Marketing, Finance—we're all interpreting it differently. There's no shared language. And without tying our data initiatives directly to business outcomes, we're chasing vanity metrics instead of real value." Tom chuckled bitterly. "So, we're running a technology circus without a ringmaster."

Richard smiled faintly. "Something like that. But we can fix it, if we stop letting technology lead and start building a data strategy that answers the right questions. It's not about more dashboards; it's about making data the connective tissue of our decisions."

The challenge was clear. If they wanted to break the cycle of failed data initiatives, they needed a radical shift, one that prioritized business-first thinking over technology obsession.

A silence hung in the air, heavy with unspoken doubts. How could a company so rich in data feel so...lost?

For Richard, the real work was just beginning.

The Data Illusion

For years, organizations have been told that data is the new oil—a precious resource that will fuel innovation and deliver a competitive edge. The promise was simple: collect data, invest in the latest AI technologies, and the magic would follow.

But the reality? Most companies drown in data while starving for insights. The illusion is seductive: If we just gather more data, insights will follow. But the harsh truth is this—data by itself doesn't drive value. What matters is how data is connected to business decisions and organizational structures that can act on those insights.

Why Data and Analytics Initiatives Fail

To understand why companies fall into this trap, the following sections break down these most common failure points:

- **The technology-first trap:** Technology alone won't save you
- **The siloed organization:** Disconnected teams and fragmented data derail progress
- **The missing link:** When business strategy ignores data
- **The conveyor belt problem:** Assembly-line thinking doesn't work in a data world
- **The data illiteracy epidemic:** Leadership doesn't always "get" the data
- **The autonomous illusion:** GenAI and agentic systems can scale misalignment instead of value

The Technology-First Trap: Why Technology Alone Won't Save You

Many organizations believe that better technology equals better outcomes. They invest in cutting-edge cloud platforms, machine learning models, and real-time analytics, but overlook the human and operational side of the equation.

If there's one thing executives love, it's shiny new technology. The promise of artificial intelligence, machine learning, and advanced analytics is irresistible. For many leaders, investing in these cutting-edge tools feels like a silver bullet. The more advanced the technology, the more advanced the business outcomes. But here's the hard truth: Technology, no matter how sophisticated, cannot fix a broken strategy. Let's break down the technology-first trap and explain why it's so common and why it fails.

Scenario 1: The Data Lake Mirage

Imagine this: A Fortune 500 retailer spends millions building a state-of-the-art data lake. Consultants promise that once the data from all systems such as point-of-sale, supply chain, and customer loyalty flows into one centralized repository, actionable insights will follow effortlessly. The CIO proudly announces, "We'll become a data-first company. Any question our business asks—this lake will answer it."

Two years later, the data lake is live, but nothing has changed. Store managers still rely on gut instinct for inventory decisions. The Marketing team complains that the data is outdated. Executives still receive monthly static reports that are hard to interpret. Despite the investment, the company sees no measurable improvement in customer satisfaction or operational efficiency.

Scenario 2: The Shiny Object Syndrome—Chasing the Latest Tech

A large retail chain invests in a cutting-edge real-time recommendation engine to provide personalized product suggestions across their online and in-store experiences. The goal is to increase cross-selling and customer engagement by leveraging the latest AI algorithms.

The technology team partners with a well-known AI vendor and spends months integrating the system. The recommendation engine is rolled out with much fanfare, and leadership is assured that advanced machine learning will deliver immediate value.

Six months after launch, customer engagement barely improves and, in some cases, conversion rates decline. Store managers report that in-store staff are overwhelmed by the complexity of integrating recommendations into their workflows. Online, customers complain that suggestions are irrelevant, offering winter coats to shoppers browsing beachwear.

An internal audit reveals the root cause: The AI model was optimized for technical sophistication, not business context. It relied on isolated clickstream data without incorporating product seasonality, customer preferences, or inventory availability. Worse, there was no clear alignment with the merchandising team, meaning the recommendations didn't reflect real-world product strategy.

Scenario 3: The Automation Mirage—When Technology Solves the Wrong Problem

A financial services firm launches a data lake to streamline access to customer data across departments. Executives envision an automated pipeline that allows real-time analysis, promising faster decision-making and improved compliance reporting.

The technology team focuses on building an infrastructure capable of handling petabytes of data deploying sophisticated tools for ingestion, transformation, and analysis. The project consumes millions in investment and spans two years.

Despite its technical success, business users find the data lake inaccessible and overly complex. Compliance teams still rely on manual reporting because critical data fields are missing or inconsistent. Meanwhile, customer-facing teams can't trust the data due to poor data quality and revert to legacy systems.

A post-mortem reveals the key failure: The project focused on technology over user needs. While the data lake was an engineering marvel, there was no process to define business requirements, ensure data integrity, or align outputs with what end users actually needed.

Why Do Companies Fall into the Technology-First Trap?

- **Fear of missing out (FOMO):** When competitors invest in AI, cloud, or real-time analytics, there's pressure to follow suit—even without a clear business case.
- **Tech-led narratives:** Vendors sell the idea that technology alone will solve business challenges, often downplaying the organizational and cultural work required.
- **Executive pressure for quick wins:** Leaders often view technology investments as a way to demonstrate progress quickly, even when foundational changes take longer to deliver real value.
- **Vendor-driven decisions:** Shiny new tools are often pitched as solutions, but without aligning to business goals, they become expensive distractions.

- **Overestimating technology's power:** Many leaders believe technology alone will solve structural and cultural challenges, when, in reality, people and processes are equally important.

Richard Marshall paced the length of the conference room, the hum of distant conversations echoing through the glass walls. It had been a month since the tense quarterly meeting, and despite his best efforts, the problem persisted—data was everywhere, but insight was nowhere. The more he dug, the clearer the pattern became: the company wasn't just suffering from a lack of technology. It was also suffering from a lack of connection.

This realization hit home during a conversation with Claire, the Head of Customer Experience. She had requested a churn analysis to identify why customers were leaving. After weeks of waiting, the Data Science team finally delivered a 40-page report, but it was useless.

"I don't understand half of it," Claire admitted, frustration creeping into her voice. "It's black box probabilities, model weights, correlation matrices. I just need to know why our high-value customers are leaving and what we can do to stop it."

Richard had seen it before: data silos creating knowledge gaps. Each function—such as Marketing, Finance, and Operations—had its own data, its own tools, and its own priorities. There was no common framework for sharing insights or aligning on business goals.

The result? Confusion, duplication, and missed opportunities. He decided to dig deeper, meeting with leaders across the organization. Each conversation revealed a new layer of dysfunction.

The Siloed Organization: Why Disconnected Teams and Fragmented Data Derail Progress

In many organizations, data lives in isolated pockets. Each department collects, analyses, and guards its information without sharing it across the business. This fragmentation leads to incomplete insights and misaligned decisions. Marketing tracks customer engagement but lacks visibility into downstream revenue. Finance focuses on cost control but misses early warning signs from customer churn. Product teams experiment with new features but remain unaware of their long-term impact on user retention. These disconnected data streams create blind spots that prevent leaders from seeing the bigger picture, resulting in missed opportunities, duplicated efforts, and decisions based on partial truths rather than holistic understanding.

The consequences of data silos extend beyond operational inefficiency; they undermine strategic initiatives. For example, a company might invest heavily in a new AI-driven customer experience while its supply chain remains blind to the resulting surge in demand. Without integrated data flows, each team optimizes for its own goals, often at the expense of the broader business strategy. Breaking down these silos requires more than technology; it demands cultural and structural change. Organizations must create cross-functional collaboration, incentivize knowledge-sharing, and implement unified data governance frameworks. Only by dismantling these barriers can companies transform fragmented data into cohesive, actionable insights that drive real business value.

Let's look at this further through the lens of three scenarios in the context of Marketing, Product, and Operations teams.

Scenario 1: The Marketing Black Hole—When Data Stays Locked in Departments

Richard's next stop was with Tom from Marketing. They were spending millions on customer acquisition campaigns, but when Richard asked which segments were driving the highest lifetime value (LTV), Tom's answer was vague.

"We track campaign engagement click-throughs and impressions, but Finance doesn't share downstream revenue data with us," Tom explained. "We can tell you who clicks, but not who buys and certainly not who stays."

In other words, Marketing was optimizing for activity, not outcomes. Without connecting their data to Finance, they were flying blind.

Richard shook his head. "So, you're telling me that Marketing is optimizing campaigns without knowing which customers actually drive profits?" Tom sighed. "Pretty much. We've asked for that data before, but Finance says their systems are separate, and it's a headache to match things up."

The silo wasn't just technological, it was cultural. Each team saw its data as proprietary. There was no shared vision for how data should flow across the business to drive decisions.

Scenario 2: The Product Blind Spot—When Teams Don't Share Insights

Later that week, Richard caught up with the Product team. They had launched a new feature six months earlier, one that was supposed to increase engagement. The data team had flagged an issue: power users were abandoning the platform after the update. But here's the catch no one in Product had heard about it.

"We track feature adoption," said Sara, the Head of Product. "But retention data sits with Customer Success. No one told us there was a problem."

The insight had been trapped in a silo. By the time anyone connected the dots, the damage had already been done. The company had lost hundreds of loyal customers, simply because no one had a 360-degree view of the data.

Scenario 3: The Inventory Mystery—When Systems Don't Talk

Perhaps the most glaring example came from the Supply Chain team. A sudden spike in backorders had blindsided the company's largest product line.

"We monitor inventory levels daily," said Mike, the Operations Director. "But we had no clue demand was surging until it was too late." Richard frowned. "But Marketing ran a promotion last quarter, shouldn't you have known?" Mike shook his head. "We didn't even know about the promotion until the orders started flooding in. Our systems don't integrate with theirs."

Siloed data meant no one saw the bigger picture. Marketing had fueled demand without informing Operations. Without access to real-time insights, supply chains were reactive instead of proactive.

Why Do Silos Exist?

As Richard reflected on these conversations, he realized that silos weren't just a technical issue; they were organizational by design. Silos form because of many reasons, with a few most common ones mentioned here:

- **Legacy structures:** Most companies organize themselves by function, not by value streams. Each department optimizes for its own KPIs, not shared business outcomes.
- **Competing incentives:** When teams are rewarded for their own performance, they don't prioritize collaboration.
- **Tool fragmentation:** Different departments adopt their own data systems without a unified architecture, making cross-functional analysis painful.
- **Lack of data governance:** Without clear policies for data sharing, critical insights remain locked within departmental walls.

The Missing Link: When Business Strategy Ignores Data

Despite the buzz around being "data-driven," many organizations treat data as a technical afterthought rather than an integral part of business strategy. Executives craft ambitious growth plans entering new markets, launching innovative products, and cutting costs, but fail to align these goals with a coherent data strategy.

The result?

Data teams operate in isolation, focusing on technology implementation rather than business outcomes. Without a clear connection to strategic priorities, data initiatives become pet projects—technically impressive but commercially irrelevant. This disconnect leads to a common paradox: Companies have more data than ever before, yet decision-makers lack the insights they need to drive tangible business value.

Consider a global retailer launching a digital transformation initiative. The leadership team envisions using AI to enhance customer experiences, yet the data strategy remains vague. Data engineers work tirelessly to modernize infrastructure, while Marketing continues to rely on outdated customer segmentation models. Meanwhile, the finance team focuses on short-term cost reductions, restricting investments in advanced analytics. With no unifying vision, these fragmented efforts fail to deliver the personalized experiences customers expect.

To bridge this gap, organizations must embed data strategy within business strategy, starting with clear, measurable objectives. Successful data leaders work alongside business stakeholders to identify how data can accelerate growth, reduce risk, and create new revenue streams, ensuring that every data initiative supports the broader strategic vision.

Let's look at this more closely with a few scenarios.

Scenario 1: The Product Launch Blind Spot

Imagine a multinational consumer electronics company preparing to launch a new line of smart home devices. The executive team is laser-focused on speed-to-market, pushing aggressive timelines to outpace competitors. However, the data team is brought in late, tasked only with generating post-launch performance dashboards rather than guiding critical pre-launch decisions. Without a robust data strategy tied to business goals, several key insights are overlooked:

- **Customer preferences:** Data from past product launches indicates that customers prioritize energy efficiency and seamless app integration, but these features are underemphasized in the new product.
- **Inventory forecasting:** Demand modeling data, which could have predicted regional buying patterns, is ignored. As a result, the company overproduces for low-demand regions while facing stockouts in high-demand areas.
- **Customer support readiness:** Data from previous launches reveals that new product categories drive a 40% spike in customer service queries—yet no proactive measures are taken to scale support.

The result? Despite advanced technology and a massive marketing push, the product launch underperforms due to poor customer satisfaction and supply chain misalignment. This failure stems from not integrating data insights into the strategic decision-making process from the outset.

Scenario 2: The Loyalty Program That Missed the Mark

A major retail bank launches an ambitious customer loyalty program to increase retention and drive cross-sell opportunities. The business strategy assumes that reward points and cash-back incentives will deepen customer engagement. However, the data team is only consulted after the program is rolled out, limiting their ability to shape its design.

Had data been integrated into the strategy, the team would have surfaced several critical insights:

- **Customer segmentation:** Data shows that younger customers prioritize fee waivers and digital experiences over reward points, yet the program only targets traditional incentives.
- **Attrition risk signals:** Historical churn data reveals that high-value clients often leave due to poor service experiences, not a lack of rewards.
- **Cross-sell opportunities:** Transaction data indicates that customers with mortgages are three times more likely to purchase insurance if they are targeted within 60 days of loan approval, but the loyalty program doesn't target this window.

Because these insights are missing from the program's design, the initiative fails to move the needle on customer retention. This misalignment underscores the danger of crafting business strategies without embedding a data-informed perspective from the beginning.

Scenario 3: The Marketing Budget Misfire

A global fashion brand is facing stagnant online sales and decides to double its digital advertising spend. Marketing leaders, under pressure to deliver quick wins, allocate millions to social media campaigns. Yet, the data science team—focused solely on web traffic reporting—has no seat at the strategy table.

If the business strategy had been informed by data, the team would have flagged key missteps:

- **Customer lifetime value (CLV) analysis:** Data reveals that repeat buyers drive 70% of profits, but the marketing spend targets first-time shoppers who rarely return.
- **Channel effectiveness:** Multi-touch attribution data indicates that email and loyalty programs yield a three times higher return than social ads, yet these channels receive minimal investment.
- **Inventory-driven marketing:** Real-time stock data suggests promoting overstocked products to reduce holding costs, but the campaign continues pushing bestsellers already in short supply.

By failing to integrate data-driven insights into the decision-making process, the company wastes millions on ineffective advertising, missing more profitable opportunities.

In each scenario, the root cause of failure is treating data as an afterthought rather than as an essential input to strategic decisions. When business strategies ignore data insights, organizations risk misallocating resources, misunderstanding customer needs, and missing growth opportunities. The solution lies in embedding data strategy into every stage of business planning, ensuring that technical capabilities align directly with business outcomes.

Why Does Business Often Ignore Data?

- **Strategy-first, data-later mindset:** Leaders define ambitious business goals but treat data as an execution layer, not a planning input.
- **Delayed data involvement:** Data teams are engaged too late in projects, missing the opportunity to influence strategic design.
- **Misaligned success metrics:** Business and data teams optimize for different outcomes (e.g., campaign impressions vs. customer lifetime value).
- **Data as a support function:** Many organizations still view data teams as service providers, not as strategic partners.
- **Short-term pressure:** Urgency to deliver quick wins often outweighs the discipline of building data-informed strategies.
- **Fragmented data ownership:** No single accountable owner ensures that data insights inconsistently feed into business decision-making.
- **Data illiteracy at the leadership level:** Executives without strong data literacy undervalue how deeply data could transform business outcomes.
- **Overreliance on gut instinct:** Business leaders often trust past experience over evidence, especially in fast-moving markets.

The Conveyor Belt Problem: Assembly-Line Thinking in a Data World

It was Richard's third trip to Warsaw that year. As the Head of Data Strategy, he was meeting his Eastern European team to discuss their progress on a new customer analytics initiative. Everything about the flight from Dublin was uneventful, until he stood by the carousel, waiting for his suitcase.

Five minutes passed. Then ten. The crowd around him thinned out as fellow travelers grabbed their bags and walked toward the exit. Still no sign of his luggage. For the first time in his life, his bag hadn't arrived.

A notification from the airline buzzed on his phone:

"We apologize for the delay in delivering your baggage. Your luggage is being traced and will be delivered to your location as soon as possible."

Annoyed but curious, Richard followed the signs to the Lost and Found counter. A customer care agent, a young woman named Katarzyna, greeted him politely. She scanned his boarding pass and, after a brief check on her computer, said, "It looks like your bag missed the transfer in Frankfurt. It should arrive on the next flight, and we'll send it to your hotel."

Sensing his frustration, she offered to show him how the system worked. Richard followed her through a restricted door and into the back room, a maze of conveyor belts and barcode scanners. The bags zipped past on different tracks; each one tagged with a unique identifier. She explained how every bag was supposed to follow a precise route from check-in, to sorting, to being loaded onto the aircraft.

"But sometimes," she said, pointing at a suitcase stuck at a junction, "one piece goes missing when the system assumes everything is fine. It's no one's fault—every part works, but the whole can still fail."

As he left the airport without his luggage, her words stayed with him. Every part works, but the whole can still fail.

The next morning, when the hotel concierge delivered his suitcase, the realization hit him. His missing luggage was no different from the data issues his team faced daily. Each data function from engineering to analytics was operating in isolation, much like the baggage system. If one part fell out of sync, it disrupted the entire chain. Yet, because each group was focused on their own task, no one saw the full picture or felt responsible for the outcome.

In many organizations, data flows like a conveyor belt passing through distinct teams responsible for different stages: data extraction, ETL (extract, transform, load) engineering, storage, analysis, and finally, business reporting. Each team focuses on their specific task, assuming that if they execute their portion correctly, the end product will be accurate and valuable. This approach mirrors the assembly-line model used in manufacturing, where every station performs a specialized function. While this method works for predictable, standardized processes like building a car, it breaks down in the dynamic world of data and analytics.

The problem lies in the fragmentation of responsibility. No one team owns the entire data product from source to strategic insight. Each group optimizes for their specific role, often without understanding how their work affects the broader business context. This leads to misaligned priorities, delayed insights, and a lack of accountability when things go wrong. For data to drive competitive advantage, organizations must move beyond linear, task-oriented thinking and foster cross-functional collaboration where every team takes ownership of the entire data lifecycle.

Richard's realization of this "conveyor belt problem" started with a simple question from the CFO during a quarterly review—"Why are our customer churn models always one step behind reality?" The more he dug, the clearer it became: the failure wasn't in any single part of the process. It was in the way the whole system was designed to treat data as a static product instead of a living asset that evolves with the business.

Let's look at the concept of the conveyor belt problem with two additional scenarios—the house construction dilemma and the car manufacturing factory.

Scenario 1: The House Construction Dilemma—When Specialists Miss the Big Picture

Richard's realization of the conveyor belt problem deepened when he began construction of his new home. Like most modern projects, the process was handled by specialized tradespeople—carpenters, plumbers, electricians, and painters—each responsible for their specific task. At first glance, the system seemed efficient. Every expert followed their predefined role, moving sequentially through the project like stations on an assembly line. However, it didn't take long for cracks to emerge, literally.

One evening, as Richard inspected the nearly finished kitchen, he noticed a problem: the plumbing for the sink didn't align with the cabinetry. The carpenter had followed the blueprint perfectly, building the cabinets to precise measurements. The plumber, working separately, installed the pipes according to his technical drawings. Neither had spoken to the other, and as a result, the sink couldn't fit without tearing out part of the new cabinetry. The problems didn't stop there. The electrician had placed power outlets based on standard guidelines, but when the kitchen appliances arrived, many didn't align with the available sockets. Meanwhile, the painter, working after everyone else, was frustrated to find unfinished patches of wall left behind by other trades, forcing them to redo portions of their work.

When Richard raised these issues with the general contractor, he got a familiar response: "The plumber did their job. The carpenter did theirs. Everyone followed their instructions."

The real problem wasn't the individual skill of each specialist; it was the lack of accountability for the end-to-end outcome. Each trade worked in isolation, focusing solely on their task without considering how their work fit together in the larger picture. No one owned the final product, the finished, functional home, which meant that misalignment was inevitable.

This experience echoed the same siloed thinking Richard had seen in data organizations. In a typical data-to-insight pipeline, teams operate like the tradespeople in a construction project:

- Data engineers gather and transform raw data.
- Analysts interpret and model the data.
- Business teams request and consume the insights.

Each group works with their own tools and priorities, but no one owns the full lifecycle—from raw data to actionable business decisions. And when these pieces don't align, the results are just as dysfunctional as Richard's half-finished kitchen.

Richard realized that data work is like building a house—without cross-functional collaboration and end-to-end accountability, even the most skilled contributors can produce a broken final product. True success comes not from isolated excellence but from integrating efforts toward a shared vision of the finished outcome.

Scenario 2: The Car Manufacturing Example—Optimizing for Efficiency, Not Value

Consider a car manufacturing plant. The assembly line is a marvel of efficiency; each station is optimized to perform one task: installing engines, fitting seats, or painting exteriors. Workers don't need to understand the full product; they focus on their piece, ensuring it meets specifications before passing it down the line. This works brilliantly for physical products where the outcome is well-defined and unchanging.

But imagine if the car's design changed mid-process, say, customers wanted electric engines instead of gasoline. The assembly line would grind to a halt. Parts already produced wouldn't fit, and every team would need to retool their processes. This is exactly what happens in data initiatives when business needs evolve but the data pipeline doesn't adapt.

In Richard's organization, the data engineers focused on optimizing infrastructure, speed, and scale. Analysts worked independently, delivering standard reports without questioning if they aligned with evolving business priorities. By the time the AI model reached the Marketing team, the market had shifted. Like an assembly line churning out outdated car models, the data team produced insights that no longer fit real-world needs.

The solution wasn't more speed; it was better collaboration. Just as modern car companies adopt modular assembly to adjust for changing consumer preferences, Richard realized his teams needed to move beyond rigid handoffs. Data work wasn't about hitting technical milestones—it was about delivering business value in real time.

In traditional assembly lines, like those found in car manufacturing, the process works because of predictability. Each station on the line is optimized to perform a specific, well-defined task such as installing the engine, fitting the seats, or painting the exterior. As long as each unit performs its function accurately, the final product—a fully functional car—emerges at the end of the process with consistent quality.

This approach works because car manufacturing involves static designs and fixed specifications. If a sedan blueprint calls for a 2.0-liter engine, the factory doesn't question whether customers suddenly want an electric motor. The assembly line assumes that the input and output are stable and unchanging. As a result, local optimization, where each station focuses only on its task, succeeds because the outcome is entirely predictable when all components are assembled correctly.

However, in data organizations, this rigid assembly-line model breaks down for three key reasons: dynamic requirements, fragmented responsibility, and a "pass-the-baton" mentality. The following sections discuss each.

Dynamic Requirements

Unlike the fixed blueprint of a car, business needs and data requirements evolve constantly. For instance, when customer behavior changes or a new market opportunity emerges, data strategies must pivot quickly. Yet, if teams operate like a car assembly line, performing fixed tasks without considering the broader picture, insights become obsolete by the time they reach decision-makers.

Imagine the source systems team only focuses on accurate data extraction without validating whether the data reflects changing customer behavior. Meanwhile, the ETL engineers prioritize pipeline performance but miss critical business context, like seasonal variations. By the time data arrives in dashboards or AI models, it's technically accurate but strategically irrelevant.

Fragmented Responsibility

On an assembly line, no single worker is responsible for the entire car. Each person owns only a slice of the process. This works because quality control occurs after assembly to catch defects. In data ecosystems, however, this fragmentation leads to a dangerous phenomenon: no one owns the entire data product.

For instance, the data architecture team may design a technically sound schema but overlook end user needs. Business analysts interpret reports without understanding how upstream ETL pipelines transform the data. When problems emerge, like a churn model failing in key markets, no one is fully accountable. Each team did their job, but no one connected the dots.

The "Pass-the-Baton" Mentality

In manufacturing, each station hands off a finished component to the next. In data work, however, teams must continuously collaborate and adjust based on real-time feedback. Treating the data lifecycle as a linear handoff process means context is lost with every transition.

For example, if the source data changes (like new customer segments emerging), downstream teams won't detect the impact until the final report misinforms leadership decisions. By then, it's too late to correct the problem, and the organization risks making flawed strategic moves.

Why Does Assembly Line Thinking Still Exist Today?

- **Linear process mindset:** Traditional business processes are designed for sequential handoffs, not the iterative, feedback-driven nature of data work.
- **Waterfall project management:** Many organizations still use rigid, waterfall-style delivery models where data is seen as a final step, not a continuous input.
- **Task specialization over end-to-end ownership:** Teams are organized by narrow tasks (e.g., ingestion, transformation, reporting) rather than owning business outcomes.
- **Misunderstanding of data lifecycle:** Leaders underestimate how often data needs to evolve, adapt, and be reinterpreted—not just processed once and forgotten.
- **Inflexible technology pipelines:** Data systems are architected like manufacturing lines—optimized for throughput, not for agility or changing business needs.
- **Success measured by output, not outcomes:** Teams are rewarded for "how much" data they process or dashboards they build, not for the value or impact of the insights.

- **Lack of product thinking in data:** Data is treated as a one-time project deliverable rather than a living product that needs constant care, iteration, and improvement.

The Data Illiteracy Epidemic: When Leadership Doesn't "Get" Data

Despite all the buzz about data-driven decision-making, a quiet but profound problem plagues many organizations—data illiteracy among leadership. This isn't about an inability to read numbers or dashboards; it's about a fundamental misunderstanding of how data works, how it flows through the organization, and most importantly, how to translate it into business value. Leaders who fail to grasp these concepts often see data as a technical concern rather than a strategic asset, leading to missed opportunities, poor decision-making, and misguided investments.

Richard Marshall encountered this problem firsthand during a quarterly business review with his company's executive team. The meeting, intended to showcase how data initiatives were transforming customer engagement, quickly spiraled into frustration. The CEO fixated on how much data the company was collecting, the CFO questioned the cost of maintaining large datasets, and the CMO demanded faster analytics—but no one asked the most critical question—"Are we using the right data to drive the right decisions?"

As the discussion unfolded, it became clear to Richard that the organization suffered from a data literacy gap at the top.

The leadership team could recite industry jargon—AI, predictive analytics, machine learning—but lacked a real understanding of how these technologies connected to their business goals. This disconnect wasn't just an intellectual gap—it was eroding the company's ability to deliver meaningful outcomes.

Richard's first real encounter with the "data as an expense" mindset came during a meeting with the CFO, where he pitched a customer segmentation initiative. The goal was to leverage advanced analytics to identify high-value customers, personalize their experiences, and ultimately increase customer retention. But as soon as Richard mentioned the investment required for data preparation and infrastructure, the CFO's expression hardened.

"This seems expensive. How much ROI can you guarantee in the next quarter?"

It wasn't the first time Richard had faced this kind of pushback. What frustrated him wasn't the question; it was the narrow view that all data investments must deliver immediate, measurable returns. The leadership team understood the output of data projects—dashboards, predictive models, and reports—but they rarely grasped the hidden work that made those outputs possible.

What many leaders fail to recognize is that most of the cost in a data initiative doesn't come from the algorithms or the fancy machine learning models—it comes from the painstaking work of acquiring, cleaning, and making data accessible. Each new use case—whether predicting customer churn or optimizing supply chains—requires specialized data preparation. And this process is rarely simple.

For example, Richard's team needed to combine customer transaction data, website clickstreams, and support logs to build the customer segmentation model. None of these datasets were stored in the same system, and each had gaps and inconsistencies that required weeks of cleaning. The work involved:

- Accessing siloed data sources across Marketing, Finance, And Customer Service.
- Reconciling different formats and definitions of "customer" across these systems.
- Manually cleaning and enriching data where automated processes failed.

By the time the data scientists had a usable dataset, the project had already consumed a month of labor-hours, and they hadn't even begun the modelling phase.

Richard tried to explain this to the CFO:

"Building a reliable segmentation model isn't like flipping a switch. It takes time to collect the right data and even longer to make sure it's accurate."

But the CFO remained fixated on the short-term balance sheet, asking: "Can't you just pull the data from the warehouse? Why does it take so long?"

What the CFO didn't understand was that while data storage is cheap, the real cost lies in curating and operationalizing data, especially when you're working across disconnected legacy systems. Each new business question requires new data and each new dataset demands time to explore, cleanse, and align.

Ironically, the organization's lack of investment in data accessibility was one of the main reasons every project felt so expensive. Without a centralized, well-governed data foundation, Richard's team was forced to start from scratch for every new request. Imagine a company where data from customer touchpoints—marketing campaigns, purchase histories, and support tickets—is scattered across ten different systems. In this environment:

- Data analysts spend 60-70% of their time locating and cleaning data rather than analyzing it.
- Data scientists waste months integrating datasets rather than experimenting with new models.
- Business leaders grow frustrated because projects take too long and produce limited insights.

This lack of data foundations creates an illusion: it seems like every new initiative is a costly, time-intensive endeavor—when the real issue is not having invested upfront in data pipelines and accessibility.

Richard knew that solving this problem required a mindset shift. Without a long-term investment in building reusable data infrastructure—automated pipelines, centralized data lakes, and clear governance—the company would stay locked in a cycle of inefficiency. But from the CFO's perspective, this kind of investment felt like a cost with no immediate ROI—a tough sell when quarterly targets loomed.

Part of the problem also lies in the invisible nature of foundational work. When leaders see a shiny dashboard or an AI-driven recommendation engine, they assume the hardest part is done by the technology. What they don't see is the hundreds of hours of effort that goes into making data usable in the first place.

Richard reflected on a recent conversation with a senior executive who had asked why simple questions—like tracking customer lifetime value—took so long to answer.

"We already have the data, why can't you just run the numbers?"

What the executive didn't understand was that while the data technically existed, it wasn't structured in a way that made analysis easy. Without investing in data discoverability, which is the ability for any team to quickly find and use accurate data, every new project became an uphill battle.

Richard knew he had to change the conversation. Instead of framing data work as a cost center, he needed leadership to see it as a long-term investment in agility and innovation. The truth was that short-term thinking—only investing in data when there's a clear, immediate business case—as limiting the company's ability to innovate. Richard believed that unless leadership grasped this reality, they would continue to see data as an operational cost, missing its true potential to drive strategic advantage. And until that mindset changed, every future initiative would feel like reinventing the wheel—with all the cost and frustration that came with it.

These scenarios underscored a harsh truth: without data literacy at the leadership level, even the best technical teams cannot succeed. When leaders view data as a black box or an isolated technical function, they inadvertently create barriers to value realization.

Richard realized that addressing data illiteracy wasn't just about training sessions or better dashboards—it required shifting the mindset of leadership.

As Richard left the meeting, he knew that fixing technology alone wouldn't solve the company's problems. Until leadership became fluent in data, every initiative would remain stuck in the same frustrating loop, with lots of activity and little value.

Why Does Data Literacy Gap Exist in Leadership?

- **Career path bias:** Many leaders rise through business, finance, or operations roles where deep data exposure is minimal, leaving a knowledge gap as they move into senior positions.
- **Legacy management models:** Traditional leadership models prioritized intuition, hierarchy, and experience over data-driven decision-making, shaping outdated mindsets.
- **Lack of formal data education:** Few executive development programs historically emphasized data literacy, analytics, or AI fluency as core leadership skills.
- **Perception that data is "technical":** Leaders often view data and analytics as technical back-office functions rather than strategic levers critical to business success.

- **Lack of accountability for data outcomes:** No clear ownership exists at the leadership level for the success or failure of data initiatives.
- **Investment gaps in data talent and training:** Organizations underinvest in building data capabilities because leadership doesn't fully grasp the long-term ROI of data literacy.

The Autonomous Illusion: When GenAI and Agentic Systems Scale Misalignment Instead of Value

As organizations begin to adopt generative AI and agentic systems, a new version of the technology first trap is emerging—one that is faster, more persuasive, and far more dangerous. Unlike traditional systems, these technologies don't just process data; they generate insights, recommend actions, and increasingly take decisions on behalf of humans.

The promise is compelling: automate knowledge work, accelerate decisions, and unlock productivity at scale. But beneath that promise lies a familiar risk. When deployed without clear business alignment, governance, or accountability, these systems don't solve organizational problems—they scale them.

If traditional data initiatives created confusion slowly, autonomous systems amplify it instantly. Let's break down the autonomous illusion, looking at why it feels like progress and why it often leads to unintended consequences.

Scenario 1: The AI Assistant That Knows Everything Except What Matters

A global bank rolls out an internal GenAI-powered assistant to help relationship managers prepare for client meetings. The system pulls data from CRM systems, transaction histories, and market feeds to generate summaries and recommended talking points.

The rollout is hailed as a success. Relationship managers save hours each week, and leadership celebrates increased productivity. But within months, cracks begin to appear.

Clients start receiving inconsistent advice. Some recommendations contradict internal risk policies. Others overlook key client preferences that were never captured in structured data.

An internal review reveals the issue: the system was optimized for completeness and speed—not relevance or compliance. It synthesized everything it could access, but lacked the contextual guardrails needed to align with the bank's advisory strategy and regulatory obligations.

The assistant didn't fail technically. It failed organizationally.

Scenario 2: The Autonomous Marketing Engine That Optimizes the Wrong Outcome

A consumer company deploys an AI-driven marketing engine designed to autonomously run campaigns, select audiences, generate content, and optimize spending in real time.

At first, the results look promising. Engagement metrics surge. Click-through rates improve. The system continuously learns and refines its approach, outperforming previous manual campaigns.

But over time, revenue growth stagnates. Customer complaints increase. High-value customers begin to disengage.

The root cause becomes clear: The system was optimizing for short-term engagement, not long-term customer value. It aggressively targeted users most likely to click, regardless of profitability or brand alignment.

Because the system operated autonomously, these decisions scaled rapidly across channels. By the time leadership intervened, the brand had already absorbed the impact.

The system did exactly what it was designed to do. The problem was that no one had clearly defined what should matter.

Scenario 3: The Agentic Workflow That No One Owns

A healthcare organization introduces agentic AI to streamline patient intake and triage processes. The system coordinates across scheduling, patient records, and diagnostic inputs, routing patients to appropriate services with minimal human intervention.

Operational efficiency improves almost immediately. Wait times drop. Administrative overhead decreases. Leadership sees this as a breakthrough in scalable care delivery.

Then, a series of misrouted cases surfaces. Patients with complex conditions are incorrectly categorized, leading to delays in treatment. Clinicians raise concerns, but they struggle to pinpoint where the decision logic resides.

The investigation reveals a deeper issue: Multiple agents were interacting across systems, each optimizing its own task, but no single owner had end-to-end accountability for outcomes.

The workflow was autonomous, but responsibility was fragmented. Efficiency had improved. Accountability had not.

Why Do Companies Fall into the Autonomous Illusion?

- **Overestimating intelligence, underestimating context:** GenAI systems are powerful at synthesizing information, but they lack an inherent understanding of business priorities, constraints, and trade-offs unless explicitly designed for them.
- **Speed over scrutiny:** Autonomous systems accelerate decision-making, reducing the natural friction where human judgment would typically intervene. Faster decisions are mistaken for better decisions.

- **Misaligned optimization goals:** AI systems optimize for what they are told to measure. When metrics are poorly defined—choosing engagement over value and efficiency over accuracy—the system scales the wrong outcomes.
- **Diffuse accountability:** As decisions move from humans to systems, ownership becomes unclear. Leaders assume the system is "handling it," while teams lack visibility into how decisions are made.
- **Hype-driven adoption:** The rapid rise of GenAI creates pressure to adopt quickly. Organizations deploy capabilities before establishing governance, guardrails, or clear business alignment.

Introducing the Leadership Algorithm: Cracking the Code to Data and AI Success

As Richard walked back to his office after his meeting with the CXOs, the frustration lingered. He knew the problem wasn't just about budgets or technology—it was about the mindset. Too many leaders viewed data as a commodity—a necessary operational function rather than a strategic asset. This misunderstanding wasn't due to a lack of intelligence or intent; it was due to a missing framework for thinking about data in a holistic, value-driven way.

Over the years, Richard had come to realize that successful organizations didn't treat data as a side project or a cost center. Instead, they built a systematic approach, an underlying algorithm for how they connected data, strategy, people, and technology. This insight led him to define what he would later call *the leadership algorithm*—a repeatable framework that data leaders could apply to align data investments with business goals and drive measurable outcomes.

At its core, the leadership algorithm is about recognizing that data success is not accidental—it is the result of intentional design and cross-functional collaboration. Much like a mathematical algorithm is a series of logical steps to solve a problem, the leadership algorithm offers a structured approach to address the core challenges that organizations face in making data work.

In the chapter ahead, we move beyond just identifying why things go wrong. It's time to examine how these invisible structures, misaligned teams, legacy hierarchies, lack of ownership, and so on sabotage even the most well-funded data programs. And more importantly, how we can measure them, fix them, and lead data organizations by design, not by accident.

CHAPTER 2

Structural Fault Lines: How to Measure What's Broken in Data Organizations

From the fifth-floor window of the Warsaw office, Richard Marshall could see the railway tracks stretching across the city, glinting under the morning summer sun as he had arrived early to prepare for a meeting with his European data and analytics leadership team.

Trains came and went with mechanical precision—each car linked, each track laid with deliberate intent. The sight reminded him of how a well-structured system could move even the heaviest loads efficiently. But unlike the railway below, his company's data operations lacked clear tracks and a cohesive direction. Despite years of pushing for data-driven transformation, Richard felt an unsettling sense of *Déjà vu*—another quarter had passed with stalled projects, missed insights, and a growing disconnect between the data teams and the business.

He poured himself a cup of coffee, reflecting on a conversation from the previous day. The CFO had asked a simple question: "Why does it take six months to answer basic business questions?"

S. Shivam, *The Leadership Algorithm*, https://doi.org/10.1007/979-8-8688-2897-3_2

Richard knew the problem wasn't with the talent or the tools—they had world-class data engineers, analysts, and AI researchers. The issue was structural. Data flowed through the organization like a maze—fragmented, delayed, and isolated. Each team optimized their piece of the puzzle, but no one owned the full picture.

Just like a railway requires aligned tracks, signals, and schedules to function smoothly, a successful data organization needs coordinated teams, defined roles, and a clear strategy. Without these, even the most advanced data platforms become like abandoned tracks, leading nowhere.

Most organizations are structured for efficiency, not insight. The traditional corporate hierarchy where data, technology, and business functions sit in separate silos was never designed for the fluid, cross-disciplinary demands of modern data and AI initiatives.

In many companies, data operates like a supply chain. Raw data is extracted, cleaned, modeled, and handed off in rigid stages. Data engineers focus on pipelines, analysts on reports, and business leaders on outcomes. This conveyor-belt structure (as Richard had come to call it) works well for repetitive, predictable tasks, but it breaks down in environments where data changes rapidly and innovation requires constant iteration.

Consider this: A marketing team needs customer insights to launch a personalized campaign. The request goes to the data team, who must locate and clean the data, which is often spread across multiple systems. Engineers focus on moving the data, while analysts interpret it. Each team does its job, but no one is responsible for ensuring that the marketing team actually gets usable insights in time. This fragmented approach results in delays, misunderstandings, and ultimately, missed opportunities.

Richard had seen this pattern repeat itself across industries—banks struggling with customer churn models, retailers failing to predict supply chain risks, and healthcare organizations unable to leverage patient data effectively.

The root cause? A misalignment between organizational design and data-driven objectives.

However, Richard had always struggled with one fundamental challenge—quantifying the impact of data decisions in a way that resonated with leadership. While he had a clear vision of how data teams should be structured, how processes should be streamlined, and how strategic changes could drive better outcomes, his arguments often hit a wall. Every time he pushed for reorganization, investment in data infrastructure, or a shift toward a product mindset, he was met with the same skepticism: "**What's the measurable impact?**"

Unlike sales figures, business deals, or product revenue, data efficiency, technical debt, and the cost of fragmentation were rarely captured in hard numbers. This gap made it difficult to justify changes or rally leadership around long-term data strategies. Over time, Richard realized that if he wanted to drive real change, he needed to bridge the gap between intuition and quantifiable evidence.

Richard's journey into quantifying data impact wasn't just driven by frustration—it was shaped by foundational experiences early in his career. Working on search systems, he learned how small changes in scoring logic could dramatically alter outcomes, making relevance measurable, tunable, and accountable. What stayed with him wasn't the complexity of the algorithms, but the discipline behind them—the ability to link logic directly to measurable impact. That clarity was often missing in enterprise data environments, where decisions were rarely quantified with the same rigor.

These early experiences shaped Richard's belief that data work could and should be measured. Over time, this evolved into a personal mission: to bring the same precision and accountability to data organizations, translating intuition into formulas and strategy into measurable design.

The following section explores the major structural problems that create these "structural faults" and those that form the base for the symptoms mentioned in Chapter 1 along with a scoring framework for each to measure them.

Fragmented Data Teams: Everyone Building Their Own Track

In many enterprise organizations, data responsibilities are scattered across multiple functions—Marketing, Finance, Operations, Risk, and other business domains. Each team builds and manages their own datasets, data models, and reporting processes, optimizing for their immediate needs. Additionally, there's often a centralized data team tasked with delivering insights across the company. However, instead of a streamlined, collaborative process, these teams work in isolation, each building their own data pipelines with little coordination.

This fragmentation leads to several issues:

- **Duplicate work:** Different teams pull the same data, clean it in their own ways, and repeat processes without leveraging shared resources and in turn end up creating multiple data marts/warehouses/platforms.
- **Inconsistent definitions:** Marketing may define "active customers" differently than Finance, leading to reporting discrepancies.
- **Bottlenecks:** Centralized data teams become overloaded with requests they don't fully understand, slowing down delivery.

When everyone is laying their own track without a shared blueprint, trains (data initiatives) are bound to run into bottlenecks and breakdowns.

To quantify the impact of fragmented data teams, where different functions (e.g. marketing, operations, finance) build and manage their own data silos, you need a formula that captures the inefficiencies and business consequences. The following framework blends technical and business metrics to measure the fragmentation impact.

Fragmentation Impact (FI)

$$FI = (D_r+D_d)\times C + I \times((T_r+T_d)\times C_t) + L$$

Where:

- **D_r:** The data redundancy rate (percentage of duplicate datasets across teams measured between 0 and 1).
- **D_d:** The data discrepancy rate (percentage of conflicting data definitions measured between 0 and 1).
- **C:** Cost of data management (annual cost of maintaining and reconciling these datasets).
- **I:** Integration complexity factor (weight assigned to the difficulty of integrating silos).
- **T_r:** Time spent on reconciling data (average hours per year spent by teams to align data).
- **T_d:** Time delays (increased time-to-insight due to fragmented systems measured in hours per year).
- **C_t:** Cost per hour of data work (average fully loaded hourly cost of the people involved in reconciliation and integration effort).
- **L:** Lost opportunity cost (revenue lost or delayed decisions due to inconsistent or inaccessible data).

Explanation:

- **$(D_r+D_d) \times C$:** Estimates how much of the annual data management cost is being consumed by redundancy and inconsistency.
- **$I \times (T_r+T_d) \times C_t$:** Converts reconciliation effort and delay into monetary cost, adjusted by how hard the environment is to integrate.

- **L:** Captures revenue loss, margin erosion, delayed action, or missed opportunity caused by fragmented data.

Example calculation from a large financial services company:

- D_r = 0.2 (1 in 5 datasets is duplicated across teams).
- D_d = 0.15 (15% of key metrics vary between teams).
- C = €200K/year (infrastructure and personnel costs for managing data).
- I = 2 (medium complexity in integrating systems—scale of 1 to 3).
- T_r = 960 hours/year (data reconciliation by cross-functional teams)
- T_d = 12 weeks/year (delay in accessing integrated data) = 480 hrs/year.
- C_t = Blended hourly cost of analysts, engineers, and managers = €65/hr.
- L = €300K/year (estimated lost revenue due to delayed insights).

  $FI = (0.2+0.15) \times 200K + 2 \times ((960+480) \times 65) + 300K$
  Fragmentation Impact (FI) = €557,200/year (approx.)
  FI Ratio = FI/Annual Data & Analytics Operating Cost

If the company's annual data and analytics operating cost is €2.5M, the FI ratio is €557,200/€2,500,000 = 22.3%. Table 2-1 explains the benchmarking levels for FI.

Table 2-1. *Benchmarking Fragmentation Impact*

Level	Rating	FI Ratio	Meaning
1	Minimal	< 5%	Fragmentation has limited business effect
2	Manageable	5% to 10%	Some friction exists but remains contained
3	Material	10% to 20%	Fragmentation is materially reducing efficiency and agility
4	Severe	20% to 35%	Fragmentation is a major drag on performance and scalability
5	Critical	> 35%	Fragmentation is structurally undermining value delivery

In the example, an impact of 22% falls into the Severe fragmentation crisis zone, meaning for the large financial services company, fragmentation is not just a technical issue—it's a strategic business risk affecting revenue, customer experience, operational efficiency, and competitive positioning

Reducing Fragmentation Impact

Try these approaches to reduce fragmentation impact:

- **Centralize critical data products:** Establish a small set of authoritative data domains and products that multiple teams can consume, reducing duplication and inconsistency.
- **Define and enforce canonical data standards:** Create shared definitions, metrics, and models across teams to eliminate conflicting interpretations and discrepancies.

- **Implement a federated data platform:** Use a common platform with standardized APIs and metadata, while allowing decentralized teams to publish and consume data easily.
- **Incentivize cross-team collaboration:** Align KPIs and rewards so that teams are recognized not just for their own outputs, but for how reusable and interoperable their data assets are.
- **Measure and reduce integration complexity:** Routinely audit integration points and refactor redundant or ad hoc pipelines to lower the cost and time of cross-team data stitching.

Note Fragmentation impact grows invisibly—compounding across systems, teams, and processes until innovation slows to a crawl. Organizations with an FI ratio above 20% annually are not just inefficient; they are structurally incapable of scaling data-driven decision-making.

The Absence of a Data Mesh: Still Running on Outdated Tracks

Traditional organizations typically follow centralized data models where a single data team is responsible for curating, cleaning, and serving data. This approach doesn't scale well in modern environments where data is distributed across cloud platforms, applications, and external APIs.

The concept of the *data mesh,* popularized by Zhamak Dehghani, offers an alternative: treating data as a product, owned and managed by the teams closest to its origin. Instead of routing all data through a central hub, the data mesh promotes domain-level ownership combined with shared standards for interoperability.

However, a data mesh is not a silver bullet.

In practice, organizations that adopt a data mesh without sufficient platform engineering, governance, and domain capability often experience new forms of fragmentation rather than solving existing ones. Ownership without accountability, or autonomy without standards, can lead to a proliferation of inconsistent, low-quality data products.

A data mesh, therefore, should not be viewed as a binary state but as a spectrum of maturity.

Failure Modes of Poor Data Mesh Adoption

When implemented without the right foundations, a data mesh can introduce:

- **Ownership without capability:** Domain teams are given responsibility for data but lack the skills, tooling, or resources to manage it effectively.
- **Proliferation of low-quality data products:** Multiple versions of "data products" emerge with inconsistent definitions and limited reuse.
- **Increased fragmentation under decentralization:** Instead of reducing silos, decentralization amplifies them.
- **Governance gaps:** Lack of standardization leads to breakdowns in trust, lineage, and compliance.

When Data Meshes Are Absent or Immature

At the other end of the spectrum, organizations that remain fully centralized face different challenges:

- Teams remain dependent on a central bottleneck for data access.
- Data accessibility is limited, slowing decision-making.
- AI initiatives stall due to lack of discoverability and integration.
- Delivery cycles become increasingly disconnected from business needs.

In both extremes, over centralization or poorly executed decentralization, the outcome is similar: reduced agility, poor data quality, and limited business impact.

Without adopting such principles:

- Teams remain isolated and cannot easily share or reuse datasets.
- Data quality suffers as no one feels responsible for maintaining clean, trustworthy data.
- AI initiatives stall because data isn't easily discoverable or integrated.

Organizations stuck on outdated, centralized tracks struggle to meet the needs of fast-moving AI projects, where agility and collaboration are crucial.

The absence of a data mesh in an organization typically results in centralized bottlenecks, poor data accessibility, slow delivery of insights, and increased operational complexity. To quantify this absence, we can measure the inefficiencies caused by relying on outdated centralized data architectures versus a decentralized, domain-driven approach that a data mesh advocates.

Real-World Indicators of Data Mesh Absence

- **Centralized gatekeepers:** Every data request must go through a small, overburdened central team.
- **Slow data delivery:** Long cycle times to produce datasets because each request is bespoke.
- **Duplicated effort:** Multiple teams re-create the same dataset with slight variations.
- **Manual handoffs:** Data moves through spreadsheets and manual processes, causing errors.
- **Low data trust:** Inconsistent data quality due to the lack of clear domain ownership.

We can define a formula to capture the degree of data mesh absence based on fragmentation, delivery time, and accessibility:

Data Mesh Absence Score (DMAS)

$$\text{DMAS} = (C \times A) + (D \times T) + (M \times R)$$

Where:

- **C:** The number of centralized data bottlenecks (how many processes rely on a single team for access)
- **A:** The data accessibility score (1 = very limited, 5 = fully self-service)
- **D:** Duplicate data products across domains (datasets created redundantly by multiple teams)
- **T:** Time-to-insight (average time to deliver new data requests in days)

- **M:** Manual data transfers (number of non-automated, point-to-point data integrations)
- **R:** Rework rate (percentage of time spent fixing/ rebuilding pipelines due to poor coordination)

Explanation:

- **Centralized bottlenecks (C × A):** Counts how many critical processes depend on a central data team and rate data accessibility:
  - 1 = Full manual access through a ticketing system.
  - 3 = Partially self-service with restrictions.
  - 5 = Fully self-service and real-time.
- **Duplicate data products (D × T):** Identifies how many overlapping datasets are created by multiple teams due to lack of shared ownership. Multiply this by the average time to deliver new data requests (in days).
- **Manual data transfers (M × R):** Counts the manual handoffs required for data movement and track the rework rate (percent of effort wasted fixing errors caused by manual or fragmented processes).

Example calculation from a large financial services company:

- C = 5 (core data workflows rely on a central data team)
- A = 2 (manual extracts required)
- D = 4 (teams maintain separate versions of customer profile data)
- T = 10 (days to fulfill cross-functional data requests)

- M = 8 (manual data movements—e.g., CSV extracts, APIs—without automation).
- R = 30% (percentage of work that is rework due to errors)

$$DMAS = (5\times2) + (4\times10) + (8\times0.3)$$

The Data Mesh Absence Score is 52.4. Table 2-2 explains the benchmarks for the Data Mesh Absence Score.

Table 2-2. *Benchmarking Data Mesh Absence*

DMAS Score	Impact Level
0–20	Strong mesh maturity: High accessibility, low bottlenecks, minimal duplication
21–50	Emerging mesh practices: Some bottlenecks and manual processes still persist
51–80	Low mesh maturity: Heavy reliance on centralized teams, redundant data work
81–120	Severe mesh absence: Organization operates in centralized, brittle pipelines with siloed domains
120+	Critical failure to mesh: Data platform is centralized, slow, and manual, and business agility is crippled

In this example, a DMAS of 52.4 indicates a low mesh maturity range, meaning that the large financial services company is still heavily centralized, facing noticeable bottlenecks, delays, and redundant efforts—and is at growing risk if data scale or demands increase.

Reducing the Data Mesh Absence Score

Try these approaches to reduce the data mesh absence score:

- **Decentralize data ownership:** Implement domain-driven data teams responsible for their datasets.
- **Enable self-service data:** Build data products with clear SLAs and user-friendly access.
- **Automate data pipelines:** Reduce manual movement by adopting automated data infrastructure.
- **Adopt data contracts:** Formalize standards between producers and consumers to reduce rework.
- **Measure data efficiency:** Regularly track duplication, bottlenecks, and delivery times.

Note The absence of data mesh principles doesn't just delay innovation, it creates technical and cultural bottlenecks that strangle an organization's ability to operate in real-time.

Conway's Law: When Organizational Design Shapes the Tracks

Melvin Conway stated that *"Any system designed by an organization will mirror the communication structure of that organization."* What began as an observation about software architecture has become one of the most powerful truths in modern data and AI leadership.

In the age of data platforms and intelligent automation, Conway's Law plays out painfully across enterprises. If an organization is siloed, the systems it builds will reflect those silos with fragmented datasets, disconnected models, and incompatible pipelines. In the AI era, Conway's Law extends even further: your AI systems will mirror your organizational boundaries. If teams are fragmented, your AI agents will be fragmented too.

This leads to:

- Multiple AI copilots answering the same customer question differently
- Separate recommendation engines competing for ownership of the same user journey
- Department-specific agents that cannot share context or hand off tasks
- Duplicate model training efforts across business units
- Inconsistent governance, risk controls, and prompt standards
- No enterprise memory layer or shared knowledge graph

The result is not enterprise intelligence; it is *artificial tribalism.* Consider how this unfolds in large enterprises:

- **Functional silos:** Marketing builds customer propensity models while Finance builds profitability models—but the definitions, metrics, and assumptions rarely align. Sales creates forecasting models, while Operations builds demand planning engines using different data foundations.
- **Tech silos:** Data engineers optimize for scale and reliability, while data scientists optimize for experimentation and model accuracy. AI engineers deploy copilots or agents without shared standards for governance, integration, or reuse.

- **Geographic silos:** Global firms often see regional teams creating duplicate data pipelines, local dashboards, and country-specific AI assistants with minimal alignment to enterprise architecture.

The Real Cost of Organizational Fragmentation

When an organization's structure is fragmented, data flows become fragmented, decision-making slows, and trust declines. AI amplifies this problem: disconnected organizations do not build connected intelligence. Instead of one coordinated brain, the enterprise creates many narrow minds.

Without deliberate operating model design, companies end up with:

- Multiple versions of the truth
- Redundant platforms and tools
- Slow cross-functional delivery
- Poor customer experiences across channels
- AI solutions that optimize locally but fail globally

Measuring Conway's Law in the context of data and AI involves assessing how an organization's structure influences its data architecture, pipelines, and product delivery. When teams are siloed or misaligned with data-driven objectives, it results in fragmented systems, inconsistent data flows, and inefficient collaboration. Quantifying this misalignment helps identify the extent to which organizational design is shaping (or distorting) the data landscape.

Conway's Law misalignment Effect (CLME)

$$\text{CLME} = (S \times P) + (D \times C) + (I \times R) + (A \times F)$$

Where:

- **S:** Number of independent teams managing similar datasets (the silo effect)
- **P:** Number of point-to-point data integrations required between those teams
- **D:** Data duplication rate (overlap in datasets across different teams)
- **C:** Complexity factor (number of transformations required to unify data)
- **I:** Number of cross-team incidents caused by misaligned processes
- **R:** Average resolution time (in hours) for these incidents
- **A:** Number of isolated AI systems/agents solving overlapping problems
- **F:** Friction factor caused by lack of interoperability, conflicting outputs, or inability of AI systems to collaborate

Example calculation from a mid-size financial institution bank:

- **(S × P):** 5 teams with 20 integrations = 100
- **(D × C):** 5 duplicate datasets × 15 transformations = 75
- **(I × R):** 5 cross-team issues × 50 hours = 75
- **(A × F):** 4 isolated AI systems × 20 friction score = 80

  CLME = 100 + 75 + 75 + 80

Conway's Law misalignment Effect (CLME) is 330. Table 2-3 shows the benchmarks for misalignment with Conway's Law.

Table 2-3. *Benchmarking Conway's Law Misalignment*

CLME	Meaning
0–100	Aligned: Minimal silos, well-coordinated teams
101–250	Moderate misalignment: Noticeable team and system inefficiencies
251–400	Significant misalignment: Structural silos creating business friction and fragmented AI initiatives
401-600	Severe misalignment: Collaboration breakdown, duplicated platforms, data chaos, conflicting AI systems
601+	Critical misalignment: System fragmentation at crisis level

A CLME of 330 indicates a significant misalignment. This suggests that the organization is already experiencing visible friction, such as:

- Slow product launches due to dependency bottlenecks
- Messy data duplication across business units
- Conflicting metrics in executive reporting
- AI assistants giving inconsistent answers
- Multiple teams rebuilding similar models
- Customer journeys fragmented across channels

Without intervention, costs rise faster than value creation.

Reducing Conway's Law Misalignment

Try these approaches to reduce any Conway's Law misalignment:

- **Reduce system fragmentation:** Implement a data mesh, assigning ownership by business domains rather than functional silos.

- **Standardize data definitions:** Create a shared data catalog and enforce governance.
- **Streamline incident resolution:** Implement cross-team SLAs and automated data quality checks.

Note Once CLME crosses ~250, friction accelerates non-linearly. Small issues snowball into large system-wide inefficiencies.

Data As an Afterthought: Building the Tracks Too Late

In many organizations, data is treated as a byproduct rather than a foundational asset. Decisions about systems, processes, and product development are made without considering how data will be captured, stored, or analyzed. Data becomes an afterthought—integrated only when issues arise.

This approach creates several downstream problems:

- **Incomplete data:** Systems may not capture critical data points, limiting analytical capabilities.
- **Data engineering debt:** Retrofitting pipelines onto legacy systems increases complexity and cost.
- **Lost opportunities:** Valuable insights are missed when data isn't designed for exploration.

If data pipelines are the railway tracks of an organization, designing them after the trains (products and processes) are running leads to derailments. Successful organizations embed data architecture into their systems from the start.

Quantifying technical debt caused by treating data as an afterthought requires measuring the inefficiencies, rework, and operational complexity introduced by poor data design.

Here's a structured framework to quantify data-related technical debt:

Data Technical Debt (DTD)

$$DTD = (R \times T) + (M \times I) + (A \times D)$$

Where:

- **R:** Rework due to poor data design (hours/month)
- **T:** Average cost per hour of data engineers/scientists (€)
- **M:** Maintenance complexity (e.g., number of custom pipelines or data transformations)
- **I:** Incidents caused by poor data (monthly/yearly)
- **A:** Accessibility delays (average hours lost due to lack of timely data access)
- **D:** Data consumers' average hourly cost (€)

Explanation:

- **Rework and manual fixes (R × T):** Measures the engineering time spent on fixing data quality issues, retrofitting pipelines, or reprocessing data.
- **Maintenance complexity (M × I):** Quantifies the number of custom scripts or manual interventions needed to maintain and operate data pipelines.
- **Accessibility delays (A × D):** Measures the time lost by data consumers (analysts, data scientists, decision-makers) due to poor data access.

Example calculation for a mid-sized company:

- **Rework (R × T):** 50 hours/month fixing legacy data × €120/hour = €6,000
- **Maintenance (M × I):** 12 custom jobs × 4 hours/month × €100/hour = €4,800
- **Accessibility delays (A × D):** 6 analysts × 10 hours/month × €80/hour = €4,800

  DTD = €6,000 + €4,800 + €4,800

The Data Technical Debt (DTD) is €15,600/month or €187,200/year. Table 2-4 shows the benchmarks for DTD.

Table 2-4. *Benchmarking DTD*

Annual DTD (€)	Interpretation
0–50,000	Healthy data foundations: Minor technical debt, good practices in place
50,000-150,000	Manageable debt: Some inefficiencies, beginning to slow down data teams
150,000-300,000	High technical debt: Debt materially impacting project velocity and costs
300,000-500,000	Severe technical debt: Major risk to scaling and delivery, constant firefighting
500,000+	Critical technical debt: Systems unsustainable, escalating cost spiral

In this example, an annual DTD €187,200/year falls under high data tech debt. Teams are losing real money and speed, and the company should actively invest in fixing core data issues, *not* just patching.

Reducing Data Tech Debt

Try these approaches to reduce DTD:

- **Shift left on data design:** Embed data engineers in product teams early to ensure data-first decisions.
- **Data mesh implementation:** Adopt domain-oriented data ownership to reduce redundant pipelines.
- **Invest in data governance:** Establish clear data stewardship roles to maintain quality and accessibility.

Note Data Technical Debt (DTD) is rarely visible in financial statements, but its compounding effect is deadly. Every hour spent fixing broken data pipelines instead of creating new insights is a silent erosion of a company's competitive advantage.

Lack of Clear Data Ownership: No One Managing the Junctions

Another common failure point is the absence of clear ownership over critical datasets. When multiple teams interact with the same data, it's easy to assume someone else is responsible for maintaining quality, privacy, and accuracy. Without clear accountability, no one manages the "junctions" where data intersects between teams.

This lack of ownership causes:

- **Data quality issues:** Errors and inconsistencies persist because no one is responsible for fixing them.
- **Access delays:** Teams must navigate bureaucratic hurdles to access the data they need.

- **Compliance risks:** Unclear ownership increases the likelihood of privacy breaches and regulatory violations.

Successful data organizations clearly define data stewardship roles and assign accountability to ensure smooth handoffs between domains. Quantifying the "lack of clear data ownership" involves measuring the operational inefficiencies, decision-making delays, and quality issues caused by unclear or absent ownership. This can be broken down into measurable factors such as error rates, access delays, time to resolve data issues, and duplicated efforts across teams.

Data Ownership Debt (DOD)

$$DOD = (I \times T) + (D \times C) + (F \times R) + (V \times S)$$

Where:

- **I:** Incident count related to unclear data ownership (monthly/yearly)
- **T:** Average time (in hours) to resolve these incidents
- **D:** Data access delays caused by unclear ownership (hours/month)
- **C:** Cost per hour for affected data consumers (analysts, decision-makers)
- **F:** Data inconsistencies or failures traced to unclear ownership (monthly/yearly)
- **R:** Average rework time (in hours) to fix these inconsistencies
- **V:** Number of duplicate or conflicting datasets across teams
- **S:** Average sync and reconciliation time (in hours) required for those datasets

Explanation:

- **Incident resolution time (I × T):** Tracks the number of data-related issues where unclear ownership caused delays and measures the average time to resolution due to cross-team confusion.
- **Access delays (D × C):** Measures the time lost because no clear owner can provide timely data access or approvals.
- **Data inconsistencies (F × R):** Counts the number of data inconsistencies (e.g., mismatched reports, out-of-sync data) caused by ambiguous ownership. Measures the time spent fixing those inconsistencies.
- **Duplicate data (V × S):** Identifies how many duplicate datasets exist because no one "owns" a single source of truth. Measures the time to synchronize and reconcile these datasets.

Example for a mid-sized company:

- **Incident resolution (I × T):** 15 incidents/month × 4 hours × €120/hour = €7,200
- **Access delays (D × C):** 10 hours/month × 6 data consumers × €80/hour = €4,800
- **Data inconsistencies (F × R):** 10 inconsistencies/month × 3 hours × €100/hour = €3,000
- **Duplicate data (V × S):** 6 duplicates × 8 hours × €100/hour = €4,800

DOD = €7,200 + €4,800 + €3,000 + €4,800 = €19,800/month

Data Ownership Debt is €237,600/year. Table 2-5 outlines the benchmarks for DOD.

Table 2-5. *Benchmarking DOD*

Annual DOD (€)	Interpretation
0–50,000	Strong ownership culture: Clear data ownership practices, minimal delays or inconsistencies
50,000-150,000	Manageable ownership gaps: Occasional confusion about ownership causing minor friction
150,000-300,000	High ownership debt: Ownership gaps are slowing delivery, increasing duplication and rework
300,000-500,000	Severe ownership debt: Frequent incidents, poor data access, trust issues among consumers
500,000+	Critical ownership breakdown: Complete fragmentation, major business risk from untrusted or delayed data

In this example, a DOD of €237,600/year indicates high ownership debt, meaning that the company has a serious, but still recoverable, level of data ownership dysfunction. Left untreated, it will escalate into major project delays, data quality distrust, and operational inefficiency.

Reducing Data Ownership Debt

Try these approaches to reduce data ownership debt:

- **Implement data domains:** Adopt a data mesh by assigning ownership to functional domains (e.g., Customer, Product, Financial).
- **Appoint data product owners:** Designate accountable individuals for critical datasets with end-to-end responsibility.

- **Enforce data SLAs:** Set clear service-level agreements (SLAs) for data delivery, quality, and access.

Note Data Ownership Debt compounds silently until it begins to choke organizational agility. When accountability for data becomes everyone's job, it quickly becomes no one's job.

While all the formulas and baselines Richard developed were critical, because they finally gave him a way to quantify the inefficiencies, fragmentation, and missed opportunities within the data organization, they addressed only half of the challenge. These models became powerful tools to communicate with senior leaders like the CIO, CFO, and CEO. For the first time, Richard could articulate in numbers how much was being lost due to siloed data teams, lack of ownership, misaligned org structures, and reactive investments. These formulas helped translate complexity into clarity, earning leadership attention and appreciation.

But as Richard quickly realized, explaining the problem isn't enough. While leadership valued the diagnostic insights, what they truly needed was a prescription—a blueprint for fixing the root issues. Without a clear pathway to minimize the challenges, the conversation would always stall at awareness, never evolving into transformation. That's where the real pivot began. Armed with the data to validate his observations and the urgency from leadership to act, Richard turned his attention toward designing a solution—one that would reimagine the very structure of how data teams are formed, governed, and aligned to business outcomes.

Richard believed the solution was to design organizations around data outcomes, not processes. This meant moving beyond functional silos and fostering cross-disciplinary teams where data, technology, and business worked together toward shared goals. His new model was inspired by product-centric thinking where small, autonomous teams own the entire lifecycle of a data product, from inception to delivery. This approach not only accelerates innovation but also improves accountability because each team is responsible for delivering business impact, not just technical outputs.

This realization led Richard to the idea of the *architect's blueprint*—a framework for organizational design in the age of data and AI. Just as an architect designs with the user experience of a building in mind, data leaders must design teams, roles, and processes to ensure that data flows smoothly across domains, insights are delivered reliably, and accountability for outcomes is structurally embedded. It's not about throwing more people at the problem or adopting the latest tool, it's about aligning structure with purpose.

The blueprint Richard advocates for is not a rigid org chart, but a flexible and dynamic design model. It accounts for domain ownership, cross-functional collaboration, decentralization where necessary (as in a data mesh), and centralized enablement for shared services like governance, architecture, and security. It emphasizes role clarity, so that each data professional—from engineers and analysts to product owners and domain leaders—knows where their responsibility starts and ends, and how that responsibility ties into value delivery.

CHAPTER 3

The Architect's Blueprint: Designing Organizations for Data Success

At the heart of the architect's blueprint are five core principles covered in this chapter, with measures against each of these principles to evaluate how far an organization is from these principles.

Data Product Teams: Organizing Around Business Outcomes, Not Activities

Most data teams are structured around capabilities—data engineering, data science, analytics, and governance—all working in silos. The blueprint flips this model by organizing around business value chains. In this structure, cross-functional data pods are aligned with strategic business domains (e.g., customer growth, risk and fraud, product optimization), ensuring accountability from data collection to decision-making.

S. Shivam, *The Leadership Algorithm*, https://doi.org/10.1007/979-8-8688-2897-3_3

Each pod acts like a mini start-up: empowered, autonomous, and responsible for delivering insights, products, or models that directly drive outcomes. Instead of handing off work between functions, the team collaborates end-to-end.

Challenges Addressed

Fragmentation, lack of ownership, and duplication of effort

Metric

Team Alignment Index (TAI)

TAI is a metric that reflects how closely data product teams are aligned to business outcomes, rather than siloed technical activities.

Formula:

$$TAI = (P_{ba} \times C_{bo} \times R_{du}) / (1 + S_{tn} + M_{sm})$$

Where:

- $\mathbf{P_{ba}}$**:** Percentage of teams aligned to business KPIs (0 to 1)
- $\mathbf{C_{bo}}$**:** Business owner collaboration score (0 to 1)
- $\mathbf{R_{du}}$**:** Percentage of backlog tied to business outcomes (0 to 1)
- $\mathbf{S_{tn}}$**:** Standalone technical initiatives/total product teams
- $\mathbf{M_{sm}}$**:** Percentage capacity spent in servicing mode (0 to 1)

Example calculation:

Let's say a data product organization has the following:

- P_{ba} = 60% of the teams have clear KPI or outcome focus, so 0.6

- C_{bo} = 0.55, based on engagement scores from business stakeholders
- R_{du} = 50% Jira backlog are tied to product KPIs or business unit needs, so 0.50
- S_{tn} = 4 standalone tech initiatives out of 8 team, so 0.5
- M_{sm} = 20%, indicating 20% of capacity spent firefighting or reactive requests

$$TAI = (0.60 \times 0.55 \times 0.5) / (1 + 0.5 + 0.3) = 0.092$$

Table 3-1 explains how to interpret the TAI score.

Table 3-1. *TAI Score Interpretation*

TAI Score	Interpretation
0.00–0.10	Low alignment: Teams operate in tech silos, poor business engagement, mostly reactive
0.11–0.25	Partial alignment: Some strategic focus, but still weighed down by technical fragmentation
0.26–0.40	Good alignment: Clear product mindset, regular collaboration, growing business outcome impact
0.41–1.00	High alignment: Deep integration with business strategy, outcome-driven backlogs, proactive roadmap

A TAI of 0.092 places the organization in the low alignment category. This suggests:

- Teams are still operating primarily in functional or technical silos rather than around business outcomes.
- Business ownership and collaboration with data teams is limited or inconsistent.

- A significant portion of backlog effort is focused on internal work rather than measurable value creation.
- Too many standalone technical initiatives are diluting focus and fragmenting priorities.
- High reactive servicing demand is reducing capacity for strategic roadmap delivery.

For a high alignment organization, with TAI = 0.51, values would look like these:

- $P_{ba} = 0.90$
- $C_{bo} = 0.88$
- $R_{du} = 0.82$
- $S_{tn} = 0.125$ (1 standalone initiative/8 teams)
- $M_{sm} = 0.15$

A company moving from 0.09 to 0.51 has not simply improved delivery. It has transformed structure, governance, prioritization, business trust, and product mindset.

Why TAI Works for This Principle

- It forces visibility into the productization of data efforts.
- It highlights when teams are stuck in ticket-driven service mode.
- It encourages embedding data teams into cross-functional business squads.
- I shows whether your data investments are truly tied to customer impact, revenue, or cost savings.

Tip High-performing data organizations do not win through more talent alone. They win by reducing organizational drag and aligning teams to measurable business outcomes.

Data Platform Teams: The Hub-and-Spoke Model: Balancing Scale with Specialization

To avoid duplication while maintaining agility, the blueprint promotes a hub-and-spoke model. A central hub team provides shared enterprise capabilities, while spoke teams embedded within business domains focus on localized execution and business outcomes.

Traditionally, the hub delivered core services such as data platforms, governance, tooling, and ML Ops. In the modern AI era, this mandate expands further. Platform teams must now provide common foundations for analytics, machine learning, and intelligent automation, including model serving infrastructure, prompt management, evaluation frameworks, agent orchestration layers, guardrail services, and reusable retrieval systems.

The hub enforces consistency, security, and standardization. The spokes deliver speed, context, and domain expertise. This dual structure maintains architectural integrity while preventing central teams from becoming bottlenecks.

Without a strong hub, each business unit risks building its own pipelines, tools, models, copilots, or AI agents creating fragmentation, duplicated cost, and governance risk. With an effective hub, teams innovate locally while scaling globally on shared foundations.

Challenges Addressed

Duplicated infrastructure, inconsistent tools, and delayed time-to-market

Metric

Platform Enablement Ratio (PER)

PER measures the effectiveness of your data platform teams in enabling decentralized teams (the "spokes") without duplicating infrastructure or effort—core to the Hub-and-Spoke model. It reflects whether the hub is truly creating enterprise leverage.

Formula:

$$PER = (N_{enabled} / N_{total}) \times (1 - R_{redundancy}) \times U_{reuse} \times A_{autonomy}$$

Where:

- **$N_{enabled}$:** The number of teams effectively enabled by the platform
- **N_{total}:** The total number of potential platform-consuming teams
- **$R_{redundancy}$:** Percent of redundant tools/platforms being built by other teams
- **U_{reuse}:** The weighted score (0–1) of reusable platform components being adopted (e.g., logging, orchestration, catalog, pipeline templates)
- **$A_{autonomy}$:** The weighted score (0–1) of user teams' self-service capability (e.g., ability to launch pipelines, deploy models, create agents, or onboard without platform team dependency)

Example calculation:

Let's take a real-world scenario of a mid-sized company with 10 data teams:

- $N_{enabled} = 8$, because eight teams are actively using the central platform
- $N_{total} = 10$

  $R_{redundancy} = 0.25$ (25% still building duplicate capabilities)
- $U_{reuse} = 0.70$ (strong reuse of shared services and templates)
- $A_{autonomy} = 0.60$ (moderate self-service maturity)

  $PER = 0.8 \times 0.75 \times 0.7 \times 0.6 = 0.252$

Table 3-2 explains how to interpret the PER score.

Table 3-2. *PER Score Interpretation*

PER Score	Interpretation
0.00–0.29	Low platform enablement: Spokes are unsupported, building custom stacks, low reuse, poor ROI on platform
0.30–0.59	Foundational platform: Partial adoption, still duplicated efforts, enablement improving
0.60–0.79	Evolving platform: Most teams use core platform, limited redundant tooling, improving autonomy
0.80–1.00	Mature platform model: High alignment, high reuse, low redundancy, excellent autonomy

A PER of 0.252 places the platform in the low platform enablement category. This suggests:

- Strong initial adoption, with most teams connected to the central platform.
- Redundant tooling and duplicate capabilities still exist across some teams.
- Reuse of shared services is improving but not yet consistent enterprise-wide.
- Self-service maturity remains moderate, with teams still relying on central support.
- Further maturity requires stronger platform capabilities, higher autonomy, and reduced duplication.

A mature PER of 0.80 usually looks like this:

- More than 95% team enablement
- Less than 5% redundant tooling
- More than 90% reuse of core components
- More than 95% self-service capability

A company moving from 0.252 to 0.80+ has not simply improved its platform. It has transformed architecture, standardization, reusability, team autonomy, delivery speed, and enterprise scalability.

Why PER Is a Great Fit for the Principle

- Reflects true leverage of a centralized platform. It's not just about building tools; it's about how many teams benefit from them.

- Helps identify if teams are reinventing the wheel (redundancy).
- Encourages productization of platform capabilities.
- Highlights how effectively the platform team scales themselves through self-service and reusable patterns.
- Can be tracked over time to demonstrate platform ROI to leadership.

Tip The best platform teams do not own every solution. They own foundations that allow others to build safely and quickly. In the AI era, the hub is no longer just a data platform—it is the enterprise engine room for scalable intelligence.

Product Thinking in Data: From Services to Sustainable Assets

Richard's blueprint introduced the mindset of data products—treating datasets, features (ML Feature Store is a great example), and insights not as one-off deliverables but as long-term, reusable assets. It shifted the mindset from projects and deliverables to long-term, evolving data products that are measured on product like qualities—usage, reliability, reusability, business value, maintainability, and so on.

Data catalogs, lineage tools, and metadata management were elevated to strategic investments. SLAs, versioning, and user feedback loops became part of product lifecycle management. If a dashboard or dataset wasn't used, it was deprecated just like bad code.

Challenges Addressed

Short-lived data artifacts, unclear lifecycles, poor data reuse, low adoption, solutions that fail to generate sustained business value, and products blocked by governance or compliance gaps

Metric

Data Product Maturity Score (DPMS)

DPMS is a weighted composite index that evaluates whether a data product is truly delivering value, not just existing technically. It combines six critical dimensions—usage, business value, reliability, maintainability, adoption and governance readiness—into a single score between 0 and 1, giving leaders a practical view of overall product health and maturity.

Formula:

$$DPMS = (w_1 \times U_s)+(w_2 \times BV_S)+(w_3 \times R_s)+(w_4 \times M_s)+(w_5 \times A_s)+(w_6 \times G_s)$$

Where:

- U_s: Usage score (0–1)—Percent of active users among target users
- BV_s: Business value score (0–1)—Composite score from business stakeholders and estimated financial impact
- R_s: Reliability score (0–1)—Precent of uptime + normalized incident rate
- M_s: Maintenance score (0–1)—Inverse of firefighting effort (higher = better maintainability)
- A_s: Adoption score (0–1)—Percent adoption across business units, teams, or key users
- G_s: Governance readiness score (0–1)—readiness to operate safely, compliantly, and transparently

- $w_1 = 0.18$: Usage score
- $w_2 = 0.25$: Business value score
- $w_3 = 0.17$: Reliability score
- $w_4 = 0.12$: Maintenance score
- $w_5 = 0.13$: Adoption score
- $w_6 = 0.15$: Governance readiness score

This weighting reflects a modern product mindset:

- **Usage (18%):** If no one uses it, it has limited relevance.
- **Business value (25%):** The strongest weighting, because value creation is the ultimate purpose of any data product.
- **Reliability (17%):** Trust and uptime are essential for sustained adoption.
- **Maintenance (12%):** Sustainable products should not consume excessive support effort.
- **Adoption (13%):** Broader organizational reach increases leverage and ROI.
- **Governance (15%):** Ensures products can scale or launch covering data lineage, access controls, model explainability, and regulatory audit trails

Example calculation for a mid-sized company:

- **Usage score:** 80% of target users actively using in the last 90 days
- **Business value score:** 0.70 stakeholders rated the product strong; value estimated
- **Reliability score:** 85% uptime with very few incidents

- **Maintenance score:** 0.6 moderate firefighting effort still happening
- **Adoption score:** 0.75 used across 3 out of 5 core business unit
- **Governance readiness score:** 0.55

  $$DPMS = (0.18\times0.80)+(0.25\times0.70)+(0.17\times0.85)+(0.20\times0.60)+(0.13\times0.75)+(0.15\times0.55) = 0.72$$

Table 3-3 explains how to interpret the DPMS score.

Table 3-3. *DPMS Score Interpretation*

DPMS Score	Interpretation
0–0.39	Immature: Ad hoc, unstable, barely usable
0.40–0.59	Foundational: Some usage, some structure, still fragile
0.60–0.79	Evolving product: Gaining adoption, value proven, maintainable
0.80–1	Mature data product: High business impact, reliable, self-sustaining

A DPMS of 0.72 places this product in the evolving product category. This suggests:

- Strong usage and growing adoption
- Clear business value already demonstrated
- Good operational reliability
- Some maintenance overhead still remains
- Further maturity possible through automation and broader rollout

Why DPMS Is a Great Fit for This Principle

- Encourages long-term ownership and product stewardship.
- Bridges the gap between tech and business via value and adoption.
- Scales well across multiple teams.
- Supports storytelling to CIOs and CDOs. "Here's our most mature product and here's our weakest."

Tip Many organizations mistake dashboards, pipelines, or datasets for products. A true data product creates repeatable value, earns user trust, and scales sustainably. If it is not used, trusted, and valuable, it is not yet a product.

Federated Governance: Enabling Autonomy Without Losing Control

In the old model, governance was either overly centralized—slow, bureaucratic, and disconnected from delivery—or fully ad hoc, creating inconsistency, unmanaged risk, and poor accountability. The blueprint balances both through federated governance. Local domain teams own policies and practices closest to the business, but operate within a central framework of enterprise standards, controls, and compliance protocols.

This creates contextual flexibility without sacrificing security, privacy, data quality, or regulatory discipline. Business-aligned data stewards work alongside engineers, analysts, and product teams to embed governance directly into the flow of work rather than treating it as a late-stage approval process.

In the AI era, governance must extend beyond data alone. Modern federated governance also includes model risk management, fairness controls, explainability standards, hallucination monitoring, prompt security, and responsible AI oversight. Domain teams may deploy models and intelligent agents locally, but within centrally defined guardrails.

The goal is simple: distributed ownership with unified trust.

Challenges Addressed

Central governance bottlenecks, non-compliance and policy breaches, poor data quality and unclear ownership, inconsistent controls across domains, unmanaged AI model risk, hallucinations, bias, and prompt injection exposure

Metric

Federated Governance Score (FGS)

FGS measures how effectively an organization balances decentralized ownership with centralized control across both data governance and AI governance. It rewards stewardship, standards adoption, and policy consistency while penalizing violations and unmanaged variation.

Formula:

$$FGS = (D_{go} \times A_{dc} \times G_{st} \times A_{ig}) / (1 + C_{vn} + V_{st})$$

Where:

- **D_{go}:** Percentage of data/AI domains with a designated owner, steward, or accountable lead
- **A_{dc}:** Percentage of domains adhering to central data contracts, quality SLAs, access policies, and approved AI operating standards

- **G_{st}:** Governance standards adoption score (0–1), based on audits, controls testing, and operating reviews
- **A_{ig}:** AI governance score (0–1), measuring maturity of model controls such as fairness checks, explainability, monitoring, prompt security, approval workflows, and human oversight
- **C_{vn}:** Normalized count of critical governance violations (privacy breaches, unauthorized access, unapproved models, severe hallucination incidents, prompt injection exposure, missing audit trails), which is critical violations/total domains
- **V_{st}:** Variance in standards across domains (0–1), measuring inconsistency in lineage, metadata, classification, controls, monitoring, or governance practices

Example calculation:

- D_{go} = 80% of data domains have active stewards
- A_{dc} = 70% follow standard contract and controls
- G_{st} = 0.85, strong audit performance
- A_{ig} = 0.65 (AI governance emerging but not mature)
- C_{vn} = 0.20, as 4 critical violations across 20 domains (4/20)
- V_{st} = 0.20, measured by scoring divergence across domains

$$FGS = (0.80 \times 0.70 \times 0.85 \times 0.65) / (1 + 0.20 + 0.20) = 0.221$$

Table 3-4 explains how to interpret the FGS score.

Table 3-4. *FGS Score Interpretation*

FGS Score	Interpretation
0.00–0.15	Fragmented and risk-prone: No ownership, inconsistent practices, frequent violations
0.16–0.35	Emerging governance: Pockets of ownership and standards, but still lots of variation
0.36–0.60	Balanced governance: Clear roles, broad adherence, rare violations
0.61–1.00	Federated at scale: Stewardship culture, automation, minimal friction between autonomy and control

An FGS of 0.221 places the organization in the emerging governance category. This suggests:

- Basic stewardship roles and ownership structures are beginning to form across domains.
- Some teams are following common standards, but adherence remains inconsistent.
- Governance processes still vary significantly between business areas.
- Critical policy violations or unmanaged risks are reducing trust and maturity.
- AI governance controls such as fairness monitoring, prompt security, or model oversight are present only in pockets.
- Further progress requires stronger accountability, automation, and more consistent enterprise-wide controls.

Example of federated at scale:

- $D_{go} = 0.95$ (95% of domains have accountable owners or stewards)
- $A_{dc} = 0.90$ (90% follow contracts, quality, access, and policy standards)
- $G_{st} = 0.92$ (strong audit and controls maturity)
- $A_{ig} = 0.88$ (mature AI governance with monitoring and safeguards)
- $C_{vn} = 0.05$ (1 critical violation across 20 domains)
- $V_{st} = 0.08$ (low standards variance across domains)

A company moving from 0.221 to 0.611 has not simply improved governance. It has transformed ownership, policy consistency, risk control, data trust, AI oversight, and scalable autonomy.

Why FGS Works for This Principle

- Captures both distributed accountability and central control
- Measures governance across data and AI systems, not just policies
- Encourages domain-level ownership through stewards and accountable leads
- Penalizes unmanaged risk and inconsistent practices
- Supports modern operating models such as a data mesh and AI product teams
- Gives leadership visibility into governance debt before regulators do

Tip Old governance models controlled access to data. Modern governance models must control access to decisions made by data and AI. The winning organizations will not choose between autonomy and control—they will architect both.

Skills Over Roles: Designing for Learning and Adaptation (with Leadership Data Fluency Stream)

Finally, the blueprint rethought team composition. Traditional titles such as report developer, BI analyst, and ETL specialist were replaced with more fluid, T-shaped professionals who combine deep expertise in one domain with broad working knowledge across adjacent disciplines.

Training academies, rotational assignments, communities of practice, and internal certification pathways were introduced to foster enterprise-wide data fluency, not just within technical teams. This shift enabled the organization to adapt as tools, priorities, and business models evolved. It also reduced dependency on hard-to-hire unicorn talent by growing capability internally.

The principle extends beyond delivery teams into leadership capability. Designing for adaptability means ensuring leaders don't merely approve data investments—they actively shape them. By embedding executive data literacy programs, formalizing value reporting, and linking sponsorship accountability to ROI, organizations close the gap between strategy and execution.

Without this, even world-class teams and platforms drift without direction.

What a T-Shaped Data Professional Looks Like

Tables 3-5 and 3-6 compare the skills of a modern data engineer versus a T-shaped analyst.

Table 3-5. *Modern Data Engineer*

Depth	Breadth	Business Translation
Data pipelines (Spark, SQL, Airflow)	Basic ML feature engineering	Explains model drivers to non-technical stakeholders
Cloud data platforms	Data contracts/schema governance	Translates latency issues into business impact
Performance optimization	Analytics tooling (Power BI/Tableau)	Communicates KPI trade-offs
DevOps/CI-CD for data	Product thinking/agile delivery	Prioritizes backlog by business value
Data quality engineering	API integration knowledge	Frames reliability in customer terms

Table 3-6. *T-Shaped Analyst*

Depth	Breadth	Business Translation Capability
SQL/dashboarding	Data modeling basics	Advises decision-makers
KPI design	Python automation	Converts insights into action plans
Visualization	Experimentation/A-B testing	Influences strategic reviews
Reporting governance	Data quality concepts	Challenges poor assumptions

Challenges Addressed

Rigid roles, skill silos, low workforce adaptability, leadership disengagement, high dependency on niche hires, and weak succession pipelines

Metrics

Skills Flexibility Index (SFI)

SFI measures workforce adaptability and cross-skilling capacity.

Formula:

$$SFI = (T_m \times R_{sk} \times U_{mob}) / (0.25 + S_{dep} + G_{gap})$$

Where all values are decimals (0–1):

- **T_m:** Multi-skilled team members ratio
- **R_{sk}:** Roles with reskilling pathways
- **U_{mob}:** Internal mobility rate based on skills
- **S_{dep}:** Single-point dependency risk
- **G_{gap}:** Skill gap score, on a scale from 0 (no gap) to 1 (severe gap)

Leadership Data Fluency Score (LDFS)

LDFS measures whether leadership can sponsor and govern data effectively.

Formula:

$$LDFS = (T_{dl} \times D_{dd} \times S_{di}) / (0.25 + LDGS)$$

Where:

- **T_{dl}:** Leaders trained in data literacy
- **D_{dd}:** Strategic decisions supported by data

- **S_{di}:** Strategic initiatives with data sponsor
- **LDGS:** Leadership data gap score

Data Skills Maturity Index (DSMI)

Weighted combination of workforce capability and leadership maturity capturing an organization's ability to adapt, cross-skill, and redeploy talent across evolving data needs, which is crucial in data organizations that thrive on rapid change and innovation along with maturity on measuring accountability at leadership level.

Formula:

$$\text{DSMI} = (0.7 \times \text{SFI}) + (0.3 \times \text{LDFS})$$

Example calculation. Imagine you're reviewing the SFI of your data organization:

- T_m = 65% of data talent is multi-skilled
- R_{sk} = 50% of roles have formal reskilling plans
- U_{mob} = 30% of people changed teams or projects via skills-based movement
- S_{dep} = 20% of key data projects are overly reliant on one individual
- G_{gap} = 0.3

  $\text{SFI} = (0.65 \times 0.50 \times 0.30) / (0.25 + 0.20 + 0.30) = 0.13$

Table 3-7 explains how to interpret the SFI score.

Table 3-7. *SFI Score Interpretation*

SFI Score	Interpretation
0.00–0.10	Rigid and role-locked: Skills are siloed, mobility is low, high risk of burnout or bottlenecks
0.11–0.22	Emerging flexibility: Some cross-functional knowledge, early signs of learning culture
0.23–0.38	Agile learning org: Skills are shared, mobility encouraged, better project adaptability
0.39–1	Adaptive talent ecosystem: Fluid, self-learning teams, talent can shift rapidly to address any data challenge

Why SFI Works for This Principle

- Emphasizes learning agility over static job descriptions.
- Recognizes how cross-skilling and mobility help in reducing single points of failure.
- Links skill development with organizational resilience and innovation.
- Encourages reskilling investments as a core part of data strategy.

For leadership data fluency score, Let's say:

- $T_{dl} =$ 0.6 (60% of leaders completed data training)
- $D_{dd} =$ 0.5 (50% of leadership decisions referenced formal data insights)
- $S_{di} =$ 0.4 (40% of execs sponsored a data-driven project)
- LDGS = 0.2 (Leadership Data Gap Score = 20%)

Then:

$$LDFS = (0.6 \times 0.5 \times 0.4) / (0.25 + 0.2) = 0.27$$

Table 3-8 explains how to interpret the LDFS score.

Table 3-8. *LDFS Score Interpretation*

LDFS Score	Interpretation
0.00 – 0.10	Legacy thinking decision model: Leadership sees data as operational reporting. Little literacy or sponsorship
0.11 – 0.22	Awareness phase: Leaders acknowledge data importance, but decisions remain intuition-led
0.23 – 0.38	Emerging data-first leadership: Data increasingly used in strategic decisions, sponsorship visible
0.39 – 1	Leadership-driven data culture: Executives actively steer transformation through data

Why LDFS Works for This Principle

- **Targets the leadership blind spot (called out in Chapter 1):** Many organizations treat data as an operational tool rather than a strategic asset because leadership lacks fluency. LDFS directly measures and addresses this.
- **Ensures top-down support for skill-building:** No reskilling initiative sustains without leadership belief and sponsorship—LDFS ensures that leaders aren't just approving budgets but are literate and actively involved.

- **Improves decision-making quality:** Leaders fluent in data are better equipped to demand, recognize, and support cross-functional, multi-skilled data teams—accelerating the shift from rigid roles to adaptive, skills-led structures.
- **Links strategy with execution:** LDFS ensures that as teams become more flexible and product-driven, leadership evolves too, creating a closed-loop alignment between strategy, governance, and capability uplift.

Hence:

$$\text{DSMI} = (0.7 \times 0.13) + (0.3 \times 0.27) = 0.17$$

Table 3-9 explains how to interpret the DSMI score.

Table 3-9. *DSMI Score Interpretation*

DSMI Score	Interpretation
0.00 – 0.12	Role-locked legacy model: Skills siloed, weak leadership fluency, low adaptability
0.13 – 0.24	Fragmented maturity: Some capability pockets, inconsistent sponsorship
0.25 – 0.39	Strong capability, scaling leadership: Operational maturity forming, leadership improving
0.40 – 1	Adaptive data enterprise: Strong workforce agility and leadership alignment

Why DSMI Works for This Principle

- **Bridges operational and leadership maturity:** By combining Skills Flexibility (SFI) and Leadership Data Fluency (LDFS), DSMI ensures that both frontline teams and decision-makers are evolving together—preventing a situation where operations modernize but leadership remains legacy-minded.
- **Highlights alignment gaps:** It quantifies the alignment (or misalignment) between how adaptable your data workforce is and how data-aware your leadership is—both essential for sustaining agile, product-oriented data teams.
- **Promotes continuous reskilling culture:** DSMI forces organizations to view skills development and leadership enablement as interconnected levers, not isolated initiatives.
- **Positions data capability as a strategic asset:** Moves the conversation from tactical hiring gaps to long-term, scalable organizational capability that underpins product thinking and federated governance models.

Tip Don't design your data organization around job titles or rare "unicorn" hires. Design around portable skills, continuous learning, and leadership fluency.

Leadership Algorithm: Principles and Quantification Summary

Table 3-10 maps the principles versus the metrics of the leadership algorithm.

Table 3-10. *Principle and Metric Mapping*

Principle	Metric
Data Product Teams Organizing around business outcomes, not activities	**Team Alignment Index (TAI)**
Data Platform Teams The hub-and-spoke model	**Platform Enablement Ratio (PER)**
Product Thinking in Data From services to sustainable assets	**Data Product Maturity Score (DPMS)**
Federated Governance Enabling autonomy without losing control	**Federated Governance Score (FGS)**
Skills Over Roles Designing for learning and adaptation	**Weighted Data Skills Maturity Index (DSMI)**

Richard stared at the table of numbers again—scores that revealed more than just gaps in process or design. They told the story of a system optimized for motion, not meaning. Each metric painted a picture of a data organization trying to run a marathon on a fractured track.

It was a clear, data-backed diagnosis—something Richard had long struggled to present convincingly in boardrooms. Now, he had a language of quantification that senior leaders could understand. CIOs could track strategic misalignments. CFOs could see inefficiencies as addressable overhead. CEOs could visualize transformation through an organizational lens, not just dashboards.

Still, numbers without action are just commentary. These formulas exposed what was wrong, but the five principles of the architect's blueprint offered the path to make it right. They weren't just philosophical ideals; they were structural interventions that could be operationalized and measured. And as these principles took root, organizing around business outcomes, designing hubs and spokes, embedding product thinking, enabling federated governance, and prioritizing skills over rigid roles—the same formulas could serve as before-and-after snapshots of progress.

When compared with the data illusion and structural faults from Chapters 1 and 2, it is very clear why and how the different symptoms of failure result in structural flaws and hence require careful consideration and assessment with a blueprint to enable a system and the leadership algorithm in action.

Table 3-11 lists the mapping themes covered in Chapters 1, 2, and 3.

Table 3-11. *Mapping Themes Discussed in Chapters 1-3*

Symptoms of Failure	**Structural Flaws**	**Architect's Blueprint Principle**
The Technology-First Trap Companies jump to buy tools, mistaking platforms for strategy.	**Data as an Afterthought** When technology precedes architecture, data becomes reactive—not foundational—resulting in tech debt.	**Product Thinking in Data** By treating data products as valuable, sustainable, evolving assets rather than one-off deliverables or by-products of technology projects. Makes data foundational, not reactive.
The Siloed Organization Disconnected teams, disconnected data.	**Fragmented Data Teams** Functional silos result in isolated pipelines, redundant logic, and mismatched goals.	**Data Product Teams** Organizing data teams around business outcomes instead of tech silos connects teams to purpose and value streams, breaking down fragmentation

(continued)

Table 3-11. (*continued*)

Symptoms of Failure	Structural Flaws	Architect's Blueprint Principle
The Missing Link Data isn't embedded in business strategy.	**Conway's Law in Action** When org charts don't reflect data value chains, misalignment becomes systemic.	**Data Platform Teams** Aligns org design with data product lifecycles through platform teams supporting domain-aligned data product teams—ensuring strategy flows through structure.
The Conveyor Belt Problem Linear handoffs break under data complexity.	**Absence of a Data Mesh** Lack of domain ownership or discoverability amplifies fragility.	**Federated Governance** Replaces linear, brittle pipelines with decentralized, domain-owned data products with clear governance guardrails—enabling flexibility with oversight.
The Data Illiteracy Epidemic Leaders don't understand data, and teams don't know how to translate insights.	**Lack of Clear Ownership** Without clear accountability, data stewardship and literacy fall apart, further alienating leadership.	**Skills over Roles** Focuses on creating adaptable, cross-functional talent pools and upskilling leadership and teams—embedding data literacy into the org's DNA.

Common Mistakes in Organizational Design for Data and AI

Even with the best intentions, many organizations stumble when redesigning for data success. Here are some of the most common traps to watch for:

- **Replicating old silos:** Teams are reorganized but still aligned to narrow functional lines, rather than business outcomes and data products.
- **Confusing platforms with products:** Building shared infrastructure is important, but unless it enables real, valuable data products, it remains just expensive plumbing.
- **Over-centralizing data governance:** In the name of control, governance becomes a bottleneck, slowing innovation instead of enabling responsible autonomy.
- **Ignoring skills and talent development:** New structures fail if people aren't reskilled to work across domains, products, and AI-driven workflows.
- **Leaving leadership behind:** Senior leaders approve new structures but remain passive, lacking the data literacy and active sponsorship needed to embed change.
- **Underestimating the cost of cultural change:** Changing reporting lines is easy; shifting mindsets toward shared ownership, product thinking, and business value orientation is much harder—and often overlooked.

- **Focusing only on technology, not operating models:** Without clear accountability, KPIs, and incentive structures for data success, even the best technology platforms underperform.

Tip Changing the structure is just the first step. Real transformation requires rethinking incentives, behaviors, and leadership habits that shape how teams interact with data every day.

By now, the new organizational blueprint was finally in place.

Richard stood by the tall windows of the Warsaw office, overlooking the web of train tracks crisscrossing in the distance—each line leading somewhere, each switch carefully engineered to prevent chaos. The data teams now mirrored that same sense of deliberate design: structured around business outcomes, supported by a strong platform backbone, guided by federated governance.

Yet despite all the hard work—the reorganizations, the governance councils, the skills transformation programs—Richard knew deep down that architecture was only part of the story. You could build the perfect train station. But if no trains ever arrived, or if they carried empty freight, what was the point?

Data, Richard realized, wasn't just an internal resource to be managed better. It was a product. A service. A driver of revenue. A strategic weapon.

And unless he and his teams learned to treat data as something the business would willingly pay for—something valuable, reliable, and consumable—the new architecture would eventually crumble under its own weight. He needed his teams to stop thinking like service providers and start thinking like product owners. He needed to turn data into an asset the business wanted, not something it tolerated.

And that meant reimagining everything—from how data was prioritized, designed, and delivered to how its value was measured.

As Richard turned back to his desk, a fresh notepad in hand, he scrawled the words that would define the next phase of his journey: "If you want to transform data's role inside a company, you don't just build pipelines—you build products."

In the next chapter, we follow Richard as he confronts the challenge of data product thinking: how to design insights, models, and platforms not just for technical brilliance, but for business impact, monetization, and growth.

Blueprints are necessary, but real change happens when leaders become builders.

CHAPTER 4

Data Products: Insights into Strategic Assets

Richard stood in front of a sprawling system architecture diagram projected onto the wall of a Dublin conference room. The image was almost comical in its complexity—dozens of arrows crisscrossing like spaghetti, connecting a patchwork of ETL jobs, dashboards, databases, and APIs. Every line represented an integration, every node, a team responsible for one slice of the pipeline.

"Where's the single source of truth?" asked Elaine, the VP of Customer Success.

Richard sighed. "There isn't one. There are five versions. It depends on who you ask."

This was the state of the analytics organization.

1. No Central Hub, Only Chaos

    Each domain had built its own version of critical datasets—customer profiles, merchant activity logs, and product hierarchies. There was no central hub of curated, governed data. Instead, each team maintained their own definitions and pipelines, leading to different departments reporting different

S. Shivam, *The Leadership Algorithm*, https://doi.org/10.1007/979-8-8688-2897-3_4

"truths" with endless reconciliation meetings before executive reviews and high engineering overhead due to point-to-point integrations that broke frequently.

2. Teams Organized by Skills, Not Outcomes

   The org structure mirrored a software factory conveyor belt where a Data Architecture team designed models, the ETL team built pipelines, the Data Science team trained models in isolation, a Scrum team of BI developers built dashboards, and each team handed off work to the next. No one owned the end-to-end product. Bottlenecks multiplied; accountability vanished.

3. Platform Teams Trapped in a Reactive Loop

   A talented group of application developers, API engineers, and frontend specialists worked tirelessly on the company's internal reporting platform. They built APIs, dashboards, and interfaces, but were order-takers, not innovators. Their roadmap was dictated by requests, not strategy.

Richard knew this structure had been optimized for cost-efficiency, not agility or impact. It might have made sense in the era of quarterly releases, but in the world of real-time analytics and AI-driven decision making, it was obsolete.

Richard drew the current architecture: endless arrows between disconnected teams. Then, beneath it, he drew a hub at the center with clean, outward-facing spokes.

"This," Richard said, pointing to the hub-and-spoke model, "is how we scale. Not with silos, but with teams that own outcomes from start to finish."

Inspired by Spotify's squad model and principles from the data mesh, Richard proposed reorganizing analytics into domain-aligned product teams:

- Each team owns a set of data products end-to-end, aligned with a business domain (e.g., Marketing, Sales, Finance, Supply Chain).
- Each team has the autonomy to design, ship, and maintain their products.
- Each team is supported by centralized platform teams for infrastructure, tooling, and security.

This created a balance of domain expertise and engineering scale. The new structure introduced two distinct layers, as outlined in Table 4-1.

Table 4-1. *The Hub-and-Spoke Transformation*

Layer	Purpose	Composition
Hub: Platform and governance team	Owns core infrastructure, standardized pipelines, governance, and reusable tooling	Senior engineers, data platform architects, DevOps, governance specialists
Spokes: End-to-end data product teams	Own a complete product or domain (Marketing, Sales, Finance, Ops, etc.) from ingestion to UI	Product owner, data engineer, data scientist, BI/UX designer, domain expert

Richard outlined five key roles every product team needed:

- **Product owner:** Owns the vision, roadmap, and adoption strategy
- **Data engineer:** Builds pipelines and optimizes infrastructure

- **Data Scientist/ML Engineer:** Develops models and algorithms
- **BI/Analytics Specialist:** Creates visualizations and reporting
- **Domain Expert/Analyst:** Provides business context and validates decisions

By embedding domain expertise, teams delivered products that resonated with actual decision makers instead of building generic outputs. This hub-and-spoke org structure is illustrated in Figure 4-1.

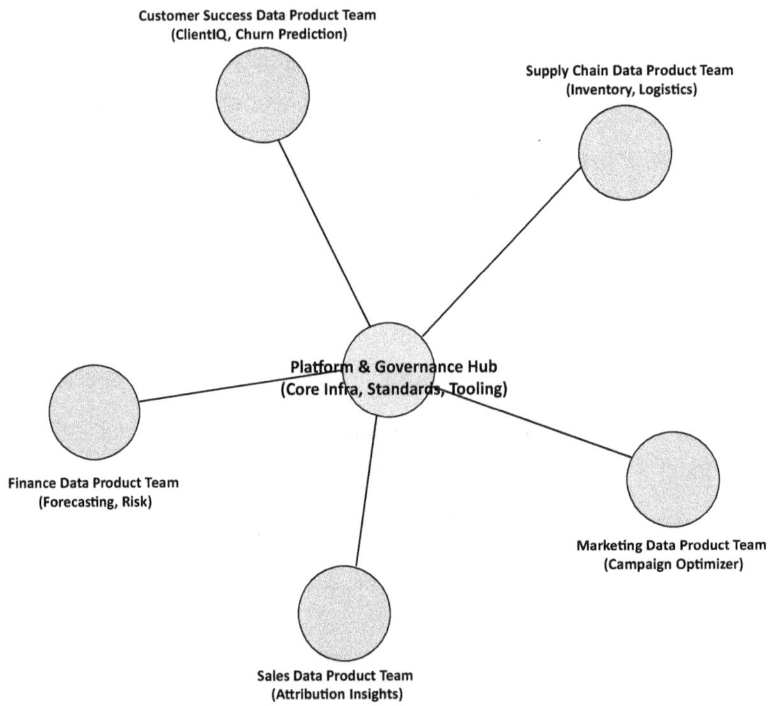

***Figure 4-1.** Hub-and-spoke org structure for data product teams*

Figure 4-1 shows how ownership shifts from silos to product teams with clear accountability and speed.

- **Yellow hub:** Central platform and governance team (core infra, standards, tooling)
- **Blue spokes:** Domain-specific, end-to-end data product teams (Marketing, Sales, Finance, Supply Chain, Customer Success)

Instead of passing tickets across five teams, a spoke team handled the full lifecycle of a product:

1. Ingesting and curating source data
2. Building models or analytics logic
3. Designing APIs or dashboards
4. Validating outcomes with business stakeholders
5. Owning adoption, SLAs, and evolution

The hub provided shared standards and tooling, while the spokes moved quickly without constantly reinventing the wheel.

Story Example: The Supply Chain Spoke

Before the re-org, supply chain analytics was a patchwork effort. It required:

- An ETL team to extract warehouse data
- A data scientist to predict restocking needs
- BI developers to visualize it
- The platform team to embed it into the internal portal

Each handoff took weeks, and there was no single owner accountable for business outcomes. After reorg:

- A Supply Chain Data Product team took ownership.

- The product owner worked directly with operations leaders.
- The engineer, scientist, and BI designer sat together, iterating daily.
- The platform team provided a standard API and template to accelerate development.

Table 4-2 outlines the benefits of this new hub-and-spoke design.

Table 4-2. *Benefits of Hub-and-Spoke Org Design*

Benefit	Description	Impact
Single source of truth	Hub manages canonical datasets, eliminating conflicting numbers.	Reduces reconciliation time; increases trust in analytics.
End-to-end accountability	Spokes own outcomes, not deliverables.	Faster delivery, clearer ownership, and better user satisfaction.
Innovation velocity	Spokes iterate quickly while hub ensures scalability.	Two to three times faster time-to-MVP.
Standardization	Hub enforces consistent pipelines, security, and governance	Reduces technical debt and rework.
Business alignment	Each spoke is tied to a domain (Marketing, Finance, Ops).	Analytics aligns directly with strategic goals.

The change was not without its challenges. Richard stood with a whiteboard marker in hand. It was Friday afternoon, and his team had gathered for what he called a "Data Product Reset" session. First, Richard had to reorient the mindset of his own analytics team. For years, they'd operated as a center of excellence—reactive, overloaded, and detached from the final user.

Many were used to thinking in tasks: "Build a churn model." "Create a Tableau dashboard." "Extract a monthly report." On the board, he drew two columns, as shown in Table 4-3.

Table 4-3. *Report vs. Data Product Comparison*

Column 1	Column 2
Dashboards, reports, extracts	Churn prediction engine, ClientIQ, merchant benchmarking API

"Let's be honest," Richard said, "which of these would you call a product?"

A few chuckles broke out. Everyone knew what he meant. They'd been churning out dashboards for years, but they didn't feel like products—they felt like outputs.

"A real product," Richard continued, "is something people depend on, trust, and return to. It has a lifecycle. It has versioning. It gets better over time."

Now, they had to start thinking like product managers:

- Who are our users?
- What problem are we solving?
- How do we measure success?
- How often are customers engaging?
- What feedback are we receiving?

Workshops were conducted. A "Data Product Playbook" was authored. And for the first time, analysts were paired directly with business counterparts to define minimum viable insight experiences instead of requirements documents.

This chapter follows Richard as he and his team embrace a new way of thinking:

- What if data wasn't an asset in potential, but a product in motion?
- What if value wasn't measured in reports shipped, but in decisions changed?
- And what if analytics could move beyond executives—into the hands of customers themselves?

What a Data Product Is and What It's Not

While still in the conference room, when moved beyond system architecture, a familiar pattern was beginning to form: KPIs, dashboards, ad hoc SQL scripts, data marts, and machine learning models—all labeled under one banner: "data products."

Richard paused. "Is that really what we're building?" he asked. "Or are we just packaging output and calling it value?"

A data product is a purpose-built, reusable, value-generating asset that uses data to solve a specific problem for a well-defined user. Like a physical or digital product, it has:

- A target user (e.g., internal analysts, business stakeholders, or external customers)
- A clear purpose (e.g., churn prediction, market segmentation, customer profitability analysis)
- Interfaces (dashboards, APIs, apps, or embeddable visualizations)

- Reliability and usability (well-documented, version-controlled, monitored)
- Evolving features (not a one-off delivery, but an iterative journey based on feedback)

A well-crafted data product moves analytics from an on-demand service to on-the-shelf intelligence.

Let's be clear on what doesn't count:

- A report manually pulled each month is not a data product.
- A table dropped into a shared folder without metadata is not a data product.
- A machine learning model trained by a team with no deployment plan is not a data product.
- A dashboard with 40 filters and no training or business context is not a data product.

If it lacks clarity, usability, ownership, or impact, it may be output, but it's not a product.

Characteristics of a True Data Product

A data product isn't defined by its tech stack or complexity. It's defined by its utility, usability, and repeatable value. Here are Richard's five rules:

1. Discoverable
    - Anyone who needs it should be able to find it.
    - Think of a Spotify-like search for datasets, APIs, or dashboards.
    - Without discoverability, products become hidden projects.

2. Usable
   - Clear interfaces, documentation, and context.
   - No more cryptic dashboards with 40 filters that only one analyst knows how to operate.
3. Trustworthy
   - Data is accurate, fresh, and explainable.
   - Users should trust that the product's numbers align with source systems.
   - Trust is earned through consistency.
4. Value-Driven
   - Each data product solves a clear business problem.
   - If no decision-maker cares about it, it's a hobby, not a product.
5. Evolvable
   - A product has a lifecycle—MVP, adoption, scale, and sunset.
   - If it doesn't change, it becomes a technical debt.

Types of Data Products

There isn't a single shape for a data product. Table 4-4 outlines a few types that Richard's team categorized.

Table 4-4. *Types of Data Products with Examples*

Type	Description	Example
Insight products	Analytical tools that support recurring decisions through reporting, trends, and KPI visibility	Store performance dashboard used weekly
ML products	Predictive or prescriptive systems that automate decisions or trigger actions	Real-time fraud detection engine
Operational data products	Trusted, reusable datasets or APIs consumed across business systems	Unified customer profile API
Embedded data products	External-facing products that expose analytics directly to customers or partners	Self-service insights for customers
Conversational/ agentic data products	AI assistants or autonomous agents that query multiple data products, synthesize answers, and initiate actions	Finance copilot that explains margin variance and opens follow-up tasks automatically
Synthetic data products	Products that generate realistic artificial data for training, testing, simulation, privacy-safe analytics, or model tuning	Synthetic claims dataset used to train fraud models without exposing real customer data

Note Think of a data product like a Spotify playlist—it doesn't create the songs, but it curates, updates, and delivers the experience consistently. You can use it again, adapt it, and share it. Reports? They're mixtapes at best.

Designing Data Products with a Value-First Mindset

Richard remembered one dashboard his Warsaw team had painstakingly built—a sophisticated multi-tab tool tracking dozens of KPIs. After months of engineering and UI work, the team finally released it with pride. But weeks later, usage metrics showed only two active users—both from his own analytics team.

Why? No one had been consulted on what decisions this tool would inform. It looked impressive, but it solved no pressing problem.

That was a turning point. Richard banned the term "dashboard request" from his team's vocabulary. From then on, everything had to start with:

- Who is the user?
- What decision is this meant to drive?
- What metric or outcome will show its success?

When Richard returned the next day, he gathered his new product team around a blank canvas—quite literally, a digital whiteboard on the large screen in the war room. "Forget features for now," he told them. "What's the value? Who are we helping? And how will we know it worked?"

Imagine a data product as a house, as illustrated in Table 4-5.

Table 4-5. *Data Product House Analogy*

Layer	Analogy	Example in Data Product
Foundation	Plumbing and wiring	Data pipelines, ETL, integrations
Rooms	Spaces where work happens	Tables, models, business logic
Doors and windows	How people access it	APIs, dashboards, apps
Furnishings	What makes it usable	Documentation, naming conventions
Curb appeal	Why people trust it	Governance, reliability SLAs

Richard often used this analogy with executives:

"You wouldn't buy a house based only on its wiring. Don't buy into data projects that don't have windows, doors, or furniture either."

Principle 1: Start with the User and the Job to Be Done

Just as successful physical products are designed around user needs, data products must start by asking:

- Who will use this?
- What decision or action will it support?
- How will it fit into their daily workflow?
- What happens if it disappears tomorrow—will anyone care?

Richard's team began mapping internal personas:

- Priya from Finance who needed margin insights per channel weekly
- David in Marketing who wanted to run "what-if" scenarios for campaign targeting
- Róisín, the product analyst, who needed funnel diagnostics at a feature level

By understanding these personas, they avoided building yet another generic dashboard that served and impressed no one.

Principle 2: Define Value Early and Often

Too often, teams define success as delivery, not impact. But product thinking demands measurable, trackable outcomes, as outlined in Table 4-6.

Table 4-6. *Example of Use Case with Product and Business Metrics*

User	Use Case	Product Metric	Business Metric
Finance	Cost optimization	Report Usage Rate	Cost per Customer Reduced
Marketing	Campaign targeting	Segmentation Accuracy	Conversion Rate Uplift
Ops	Inventory prediction	Forecast Error Reduction	Stockout Percent

Note Richard's Rule: If a data product doesn't have at least one user behavior and one business outcome metric, it's not ready for delivery.

Principle 3: Think Modular, Not Monolithic

Building data products doesn't mean bundling everything into one massive dashboard. Think modular building blocks:

- Core datasets with versioning and lineage
- APIs exposing data to systems
- Notebooks or templates that allow reproducible analysis
- UI views that can plug into tools like Power BI, Tableau, or custom apps

This modularity not only improves reusability but also enables faster iteration and clearer ownership—both vital for scaling data products across teams.

Principle 4: Define Product Lifecycle and Ownership

Too often, data initiatives die in handover—a dashboard is built, demoed, and forgotten. Richard introduced a Data Product Lifecycle Framework with four clear stages, outlined in Table 4-7.

Table 4-7. *Example of Product Lifecycle*

Stage	Goal	Timeframe	Example Criteria
Prototype	Validate concept quickly	< 4 weeks	Basic model/demo, early user testing
MVP	Ship first usable version	< 3 months	Live product with core features
Adopted	Drive adoption and impact	Ongoing	Monitored usage, SLA in place
Evolve/retire	Iterate or sunset	Annual review	ROI analysis, upgrade path, or deprecation

Each stage had a product owner, technical lead, and business stakeholder. Nothing was delivered without a clear owner.

"We don't release dashboards anymore," Richard told his team.

"We release products—and someone's on the hook to keep them useful."

Principle 5: Discoverability, Documentation, and Support

In Dublin, one of the largest blockers was finding what already existed. Different teams often rebuilt similar reports or pipelines because no one knew what was available.

To solve this:

- The team introduced a data product catalog, searchable by team, business metric, or persona.
- Each product had a one-pager—what it does, who owns it, last update, dependencies, and KPIs.
- An internal Slack bot allowed people to ask, "Is there a product for X?" and get matched with the candidates.

Within two quarters, redundant work dropped by 35%, and user satisfaction on analytics requests improved by 40%.

Principle 6: Monitor Usage and Measure Value

Inspired by how SaaS companies monitor feature adoption, Richard asked for telemetry on:

- Query volumes
- API hits
- Report usage
- NPS-style feedback on value delivered

They began publishing a monthly data product scorecard, highlighting:

- Top-used products
- Sunset candidates
- Business impact stories (e.g., "This product helped save €400K in supply chain costs last quarter")

This visibility built trust and funding.

Principle 7: Automation As a Scaling Engine

Manual work was the enemy of scale. Richard invested heavily in automation:

- **Data product templates:** Standardized API and dashboard skeletons, so teams didn't reinvent design patterns.
- **CI/CD across analytics, MLOps, and AIOps:** Automated deployment pipelines for dashboards and data products; model training, validation gates, drift monitoring, and retraining workflows for ML products, plus prompt versioning, evaluation harness runs, guardrail testing, and release controls for AI agents and copilots.
- **Auto-remediation and self-healing ops:** Failed pipelines restarted automatically, broken dashboard releases rolled back, stale models retrained, and unsafe agent behaviors disabled or escalated without waiting for manual intervention.
- **Reusable governance automation:** Access controls, lineage capture, audit logging, policy checks, and compliance evidence were embedded into delivery pipelines by default rather than added later.

- **Telemetry:** Every product automatically logged usage stats, reliability metrics, and incident data to feed into DPMS scoring.

This automation allowed Richard's small data engineering group to support three times as many products without burning out.

Breaking the Monetization Maze: Turning Data Products into Revenue Engines

Richard's shift toward treating data as a product created energy across the company—but with it came a new challenge he had never fully confronted: monetization.

For years, analytics had been viewed as a cost center—important, yes, but rarely measured in terms of revenue potential. Now, armed with a product mindset, Richard wanted to flip that narrative. He wanted data to earn its seat at the revenue table. And like any maze worth navigating, monetization brought both opportunity and confusion.

Internally, people asked: "Can we really sell data?"

Externally, customers asked: "What insights do you have that we don't?"

Richard realized quickly: the question wasn't about selling data. It was about productizing intelligence

Step 1: Identify External Use Cases, Where Real Value Lives

Instead of waiting for executives to hand him a list of monetizable ideas, Richard embedded his team inside go-to-market functions—customer success, product, partnerships, and compliance. The goal wasn't to pitch data products. It was to listen.

Within weeks, three high-potential patterns emerged:

- **SME behavioral segmentation:** Smaller businesses had limited analytics maturity. They wanted to understand their customer behavior but lacked the resources to build it.
- **Fraud anomaly detection API:** Fraud patterns seen across thousands of customers could be turned into a pre-trained model offered as a risk-scoring API.
- **Embedded benchmarking dashboards:** Retailers wanted to compare themselves with peers—conversion rates, refund patterns, regional demand trends—using anonymized aggregated data.

These weren't dashboards. They were exportable intelligence, shaped, governed, packaged, and ready to integrate.

"If the insights already exist," Richard told his team, "then we're not creating a new asset, we're simply turning intelligence into a product."

Step 2: Treat Data as a Service (DaaS): The Infrastructure of Monetization

Data monetization can't run on ad hoc reports. It needs platform-level thinking. Richard introduced a Data-as-a-Service (DaaS) layer exposing company intelligence as modular components:

- Customer metrics and behavioral clusters
- Product usage signals
- Real-time risk scores
- Sector and geographic trend analytics

Each DaaS component came with:

- SLAs (uptime, latency, concurrency)
- Pricing options (per-call billing, monthly access, enterprise licensing)
- Security classifications
- Clear ownership models

The first pilot targeted channel partners who serviced long-tail merchants. Within six months, the risk benchmarking API became a premium add-on, unlocking a new revenue stream that required zero new data collection.

The intelligence was always there; it had simply never been packaged.

Step 3: Build Operational and Legal Guardrails Monetization Without Missteps

As soon as revenue entered the picture, so did risk. Customers, regulators, and partners all asked different variations of the same question:

> *"If you're monetizing insights, how do we know they're compliant, explainable, and fair?"*

Richard partnered with legal, risk, and security to define a Data Monetization Framework, including:

- Classification tiers: internal, sharable, monetizable
- Consent and opt-out flows
- Audit logs for every AI-powered decision
- Bias assessment standards
- Explainability requirements for customer-facing intelligence

This framework protected trust—internally, externally, and regulatorily. Richard knew that one mistake in monetized insights could cost more than all the revenue it brought in.

Step 4: Report ROI, Not Just Usage

To earn continued support from the C-suite, Richard published quarterly business impact updates that highlighted:

- Licensing revenue generated from APIs
- Uptake of premium analytics tiers
- Conversion improvements tied to data-driven recommendations
- Retention lifts from intelligence-driven experiences

These weren't vanity metrics like "number of dashboards created." These were board-level numbers, tied directly to business outcomes.

Richard introduced a simple litmus test before allowing any insight to be called a "data product." The four stages of this litmus test are outlined in Table 4-8.

Table 4-8. *Stages of a Data Product*

Stage	Goal	Timeframe	Example Criteria
Prototype	Validate concept quickly	< 4 weeks	Basic model/demo, early user testing
MVP	Ship first usable version	< 3 months	Live product with core features
Adopted	Drive adoption and impact	Ongoing	Monitored usage, SLA in place
Evolve/retire	Iterate or sunset	Annual review	ROI analysis, upgrade path or deprecation

Intelligence without a problem, without ownership, or without measurable impact was not a product. Table 4-9 shows the pertinent questions to ask.

Table 4-9. *Data Product Questionnaire*

Question	Why It Matters
Does this solve a real decision problem?	Avoids vanity dashboards
Is it discoverable and documented?	Eliminates duplication
Does it have clear ownership?	Prevents "who fixes this?" chaos
Can engagement and value be measured?	Proves ROI
Does it have a roadmap?	Signals maturity and evolution

If any answer was "no," Richard shut it down.

Even after early monetization wins, Richard noticed a pattern—teams repeatedly assumed data product monetization was a technology challenge—more models, more APIs, and more dashboards.

But the hard part wasn't tech. The hard part was business strategy.

- Who is the customer?
- What pain point does this intelligence solve?
- How will it be priced, sold, packaged, and renewed?
- Who owns customer support?
- How will we measure success?

Monetization wasn't about algorithms. It was about commercialization.

The Real Economics of Data Products

As organizations mature in their data product journey, monetization can no longer be approached as a technical exercise. It requires understanding the economic models, pricing strategies, and value mechanics that determine whether a data product becomes a revenue engine or remains an internal tool. Monetization is ultimately a question of business design, not model performance.

This section provides a structured perspective on the real economics behind data products, and the commercial levers leaders must consider.

Direct vs. Indirect Revenue

Data products create value in more than one way. Organizations often focus solely on direct monetization, but the majority of economic impact comes from indirect business value.

Direct Revenue Models

These represent explicit monetization structures where the organization charges for insight or intelligence:

- Subscription-based analytics products
- Pay-per-use or consumption-based APIs
- Licensing of proprietary datasets
- Premium AI-driven product tiers
- White-labeled or OEM intelligence modules

Direct revenue is measurable and predictable, and it forms the financial backbone of external-facing data products.

Indirect Revenue Models

Indirect value is generated when intelligence improves operational or customer outcomes, even if customers do not pay directly for the insights:

- Reduced churn through predictive recommendations
- Lower operational costs via automation and anomaly detection
- Better pricing decisions informed by predictive models
- Improved fraud loss ratios
- Higher product adoption and engagement
- Reduced time-to-decision for internal teams

In mature organizations, indirect value often outweighs direct revenue. Therefore, economic assessments must include both dimensions.

Pricing Models for Data and AI Products

Pricing determines market adoption as much as features or technical performance. Unlike traditional software, pricing data and AI products requires aligning commercial structure with the nature of intelligence being delivered. Common pricing models include the following prices.

Consumption-based pricing:

- Charges per API call, prediction request, or data pull
- Suitable for developer-facing and technical products
- Scales with usage

Tiered/subscription pricing:

- Different levels of analytics or AI capability
- Works well for SaaS models
- Encourages predictable ARR (Annual Recurring Revenue)

Value-based pricing:

- Pricing tied to customer value (e.g., cost savings or revenue generated)
- Requires strong measurement frameworks
- Ideal for high-impact AI features

Outcome-based pricing:

- "Pay if fraud reduces by X%"
- "Pay if credit losses drop by Y%"
- Aligns incentives but demands transparency and explainability

Embedded upsell pricing:

- Intelligence added as a premium feature inside existing products
- Often the fastest route to monetization
- Requires no separate sales motion

Selecting the right pricing approach depends on customer maturity, regulatory constraints, industry norms, and how measurable the product's impact is.

Value Articulation: Communicating Economic Impact

Even when a data product performs well technically, monetization fails if stakeholders cannot understand its economic contribution. Value articulation requires translating model outputs into business outcomes.

Examples of effective value articulation include:

- From model accuracy to dollar impact
- From prediction to cost avoided
- From risk score to claims prevented
- From customer segment to conversion uplift
- From forecast accuracy to optimized planning decisions

Clear articulation includes:

- Before/after comparisons
- Scenario modeling (best case, typical case, worst case)
- Financial proxies to estimate economic benefit
- Operational KPIs tied directly to product outputs

Data products sell on outcomes, not algorithms.

Monetization As a Business Model Challenge

Technology alone does not determine monetization potential. Most data products fail commercially due to gaps in business model design, not technical capability.

Common causes of monetization failure include:

- Unclear customer segment or value proposition
- Lack of go-to-market motion (sales enablement, packaging, support)
- Misaligned incentives across product, sales, partner, and compliance teams
- Pricing models that do not reflect the buyer's willingness to pay
- Absence of a defined renewal or upsell path
- Poor integration into customer workflows or internal systems
- Lack of ownership for sustaining and enhancing the product

Therefore, successful monetization requires:

- A clear customer problem
- A defined commercialization model
- A structured pricing strategy
- Embedded sales and support processes
- Continuous measurement and feedback loops

In mature organizations, each data product is treated like a micro-business with its own customers, economics, and roadmap.

Why Understanding Economics Matters Before Scaling

As Richard's organization expanded its data product portfolio, it became clear that not all products had equal commercial viability. Some were strong internal enablers but weak external offerings. Others had strong external potential but required significant regulatory preparation.

Understanding the economics of each data product became essential for:

- Prioritization
- Investment allocation
- Pricing decisions
- Product-market fit assessment
- Go-to-market strategy
- Long-term sustainability

This naturally led to the realization that the organization needed a consistent method to evaluate data product maturity—not only technical maturity, but economic and adoption maturity as well.

The next step was to formalize this understanding into a framework that could guide teams across discovery, design, commercialization, and scale.

This need gave rise to the Data Product Maturity Scale (DPMS).

Example: BI as a Service—A Monetizable Data Product in Practice

To illustrate how the economics and monetization mechanics of data products come together in practice, consider a product like BI as a Service (BIaaS)—a packaged analytics experience delivered to external clients as a subscription offering.

BIaaS is one of the most accessible forms of data product monetization because it leverages intelligence that the organization already possesses and presents it as a curated, high-value service.

What the Product Is

BI as a Service is a ready-made analytics suite that provides customers with:

- Standardized dashboards and insights
- Trend analysis using aggregated industry data
- Predictive indicators such as churn and risk scoring
- Benchmarks comparing a customer's performance to peers
- Automated anomaly detection and alerts
- Embedded reports inside existing customer tools or portals

The value lies not in raw data, but in curated intelligence: Clean, governed, cross-customer insights packaged into a product that requires no setup, no analytics skill, and no engineering maintenance from the customer.

The Customer Value Proposition

BIaaS creates value for customers across four dimensions:

a. **Operational Efficiency**

   Customers no longer have to build or maintain analytics systems. The BIaaS platform saves cost, time, and internal engineering effort.

b. **Better Decision Making**

The dashboards translate raw data into:

- Performance indicators
- Market trends
- Forecasts
- Recommendations

Customers gain the ability to make smarter, faster decisions backed by trusted intelligence.

c. **Industry and Peer Benchmarking**

Aggregated, anonymized insights reveal how customers compare to:

- Similar businesses
- Regional cohorts
- Market segments
- Historical patterns

This alone often justifies the premium tier.

d. **Predictive and Prescriptive Intelligence**

AI-augmented features such as:

- Predictive demand curves
- Customer churn risk
- Payment or fraud risk
- Inventory optimization recommendations

These act as digital "decision advisors," increasing product stickiness and customer trust.

Revenue Model and Pricing Strategy

BIaaS allows multiple monetization paths.

Subscription tiers include:

- **Basic:** Operational dashboards, standard KPIs
- **Growth:** Benchmarking, automated insights
- **Advanced:** Predictive models, trend forecasting
- **Enterprise:** Custom integrations, API access, white labeling

Before setting any price, a product manager must understand that pricing is the output of a structured process—not a random decision.

For BIaaS, determining the price requires assessing cost, value, market dynamics, and scalability.

The following section is a realistic breakdown, mirroring how pricing is done in enterprise B2B SaaS and data products.

Build vs. Buy: The Foundational Decision

Before pricing, PMs must choose between these two options.

Option A: Build In-House

Costs include:

- Engineering effort (backend, frontend, ML)
- Data engineering and pipeline maintenance
- Visualization layer or UI framework
- DevOps and monitoring
- Security and compliance
- Ongoing feature development

Advantages:

- Full control
- Customization
- Long-term margin expansion

Challenges:

- High upfront investment
- Longer time to market

Option B: Buy and White-Label

Examples: Sisense OEM, Looker embedded analytics, PowerBI embedded, Tableau OEM.

Costs include:

- Licensing cost per client/per model
- Integration and customization
- OEM support agreement
- Potential revenue-sharing

Advantages:

- Faster time to market
- Lower engineering burden
- Enterprise-grade stability

Challenges:

- Lower margins
- Dependency on vendor roadmap

Pricing cannot be set until the PM knows whether they're selling their own technology or someone else's. This is Step 1.

Cost Modeling: Understanding the Economics Before Pricing

A PM next builds a cost-to-serve model. These are the typical cost components:

a. **Cloud Compute and Storage**

- Data warehouse compute (Snowflake/BigQuery/Redshift)
- ETL/ELT pipeline runs
- Refresh frequency (daily/hourly/real-time)
- Data storage growth
- Caching and query acceleration

A realistic anchor cost might be:

- €0.60–€1.50 per dashboard refresh per client per day
- Or €200–€1,000/month per client depending on usage

b. **Licensing Costs (for OEM or Embedded Solutions)**
Examples:

- €10–€30/user/month for embedded BI
- Revenue-share models (5–20%)
- OEM minimum contract commitments

c. **Support and Customer Success**

- Tier 1 and Tier 2 support
- Documentation
- Onboarding/implementation time

  Often, €300–€800 per client per month is allocated internally.

d. **Engineering and Product Maintenance**

    - Enhancements to dashboards
    - New predictive models
    - Security/compliance updates
    - SLA monitoring

- This is allocated as an amortized cost (e.g., €X cost per quarter ÷ expected number of clients)

Cost-to-Serve Formula (Typical PM Model)
A BIaaS PM usually works with something like this:

> Total Monthly Cost per Client = Cloud Compute + OEM Licensing + Support Cost + Allocated Engineering Cost

Only after this is understood can they set a floor price—the minimum price at which you cannot lose money.

Break-Even Modeling: "How Many Clients Do We Need?"

PMs model pricing by calculating the break-even point:

Step 1: Identify fixed costs

- Initial build (e.g., €500k)
- Annual engineering support (€300k)
- Licensing minimums (€120k/year)

Step 2: Identify variable cost per customer
Example: €350/month per client
Step 3: Determine pricing scenarios**
Example: €500, €1,000, €2,000 per month

Step 4: Calculate break-even clients

PMs now create a table, as shown in Table 4-10.

Table 4-10. *Product Break-Even Simulation*

Monthly Price	Monthly Margin	Clients Needed to Break Even
€500	€150	3,333
€1,000	€650	769
€2,000	€1,650	303
€5,000	€4,650	108

This quickly shows whether cheaper tiers are sustainable or whether higher-value features must exist for viability.

TAM, SAM, SOM: Revenue Potential Based on Market Size

PMs must confirm that the addressable market supports the pricing.

Total Addressable Market (TAM)

Total number of possible clients × annual price

Example: 50,000 SMEs × €2,000/year = €100M TAM

Serviceable Available Market (SAM)

Segment that fits product capabilities today

Example: 10,000 digitally mature SMEs = €20M SAM

Serviceable Obtainable Market (SOM)

Portion realistically obtainable in two to three years

Example: 3,000 clients = €6M SOM

This ensures that pricing aligns not only with cost but also market demand.

Competitive Benchmarking

A PM reviews competitor products and collates the information shown in Table 4-11.

Table 4-11. *BI Product Competitive Benchmarking*

Competitor	Price	Differentiator	Gap
Looker OEM	€30/user/month	Strong BI	Hard to customize
Tableau embedded	€25/user/month	Ease of embedding	Expensive for scale
PowerBI embedded	€5/user/month	Cost-effective	Limited extensibility
Niche BI vendors	€500–€2,000/month	Industry focus	Limited AI

This tells the PM whether their own BIaaS should be:

- Premium (high differentiation)
- Mid-market (value-focused)
- Entry-level (volume focused)

Willingness-to-Pay (WTP): Customer Validation

Before finalizing the price, PMs test the market with:

- Surveys
- Interviews
- Conjoint analysis
- Pricing experiments
- A/B tests with tiers

Questions include:

- "What would make this product worth €1,000/month to you?"
- "How does this compare to building your own analytics internally?"
- "Would predictive features justify a higher tier?"

WTP is the final guardrail.

Translating All Inputs into a Pricing Strategy

Only now can a PM combine:

- Cost-to-serve
- Market size
- Competitive pricing
- Willingness-to-pay
- Break-even analysis
- Pricing psychology (round numbers, parities, bundles)

This produces the actual pricing model; they emerge from structured product economics.

Why This Matters for Data Leaders

This example shows that:

- Pricing data products is analytical, not creative.
- Data product monetization requires business acumen.
- Cost and value both determine price.

- Technical excellence is irrelevant if pricing is poor.
- Data product teams must partner with Finance, Sales, and Product Marketing.

Why BI As a Service Works As a Monetizable Product

Several characteristics make BIaaS a strong example of a scalable data product:

- **High reusability:** Same engine, many customers
- **Standardization:** Minimal customization required
- **Repeatable onboarding:** Plug-and-play integrations
- **Measurable value:** Easily tied to KPIs like revenue, churn, and efficiency
- **Low marginal cost:** Once built, each additional customer costs very little
- **Expandable:** Add new dashboards, new AI models, and new alerts over time
- **Ecosystem-friendly:** Can integrate via APIs into partner platforms

BIaaS represents the sweet spot where data + intelligence + productization + packaging + pricing = predictable revenue.

Where BIaaS Often Fails (and Why DPMS Matters Next)

Despite its potential, BIaaS can fail if:

- The insights are not standardized.
- Data quality varies across sources.
- Benchmarking logic is inconsistent.

- The product lacks documentation or governance.
- There is no defined owner for enhancements.
- Support teams cannot explain the intelligence behind the dashboards.
- Adoption metrics are not tracked.
- Stakeholders treat it like a project, not a product.

This is exactly why organizations need a maturity model. A framework like DPMS (Data Product Maturity Scale) becomes essential to assess:

- Technical robustness
- Product experience quality
- Operational readiness
- Governance and trust
- Commercial viability
- Adoption and sustainability

BIaaS becomes the perfect bridge. It shows what a mature data product should look like, and what breaks when maturity is lacking.

The DPMS Framework: Measuring Data Product Maturity

As organizations begin treating data assets as monetizable products rather than analytical outputs, one challenge becomes immediately clear—not every insight, model, or dashboard is ready to operate as a product. Some are stable, scalable, and commercially viable. Others remain prototypes, one-off analyses, or fragile experiments.

To move from experimentation to repeatable value creation, leaders need a consistent, objective method for evaluating the maturity of each data product. Pricing, commercialization, adoption, reliability, and long-term sustainability all depend on knowing where a product stands today and what it needs to advance.

This is where the Data Product Maturity Score (DPMS) becomes essential.

DPMS provides a standardized way to evaluate a data product's readiness across five dimensions—usage, business value, reliability, maintenance effort, and adoption. It gives leaders a single, comparable metric to guide:

- Investment decisions
- Roadmap prioritization
- Commercialization readiness
- Go-to-market timing
- Sunset vs. scale discussions
- Engineering and product resourcing

DPMS reframes maturity not as a subjective judgment but as a measurable progression. See Table 4-12.

Table 4-12. *DPMS Score Referenced from Table 3-3*

DPMS Score	Interpretation
0–0.39	Immature: Ad hoc, unstable, barely usable
0.40–0.59	Foundational: Some usage, some structure, still fragile
0.60–0.79	Evolving product: Gaining adoption, value proven, maintainable
0.80–1	Mature data product: High business impact, reliable, self-sustaining

This simple interpretation framework gives leaders the ability to quickly classify and communicate product readiness.

Example: Scoring a Churn Prediction API

To demonstrate the forward-looking nature of DPMS, consider a Churn Prediction API being evaluated for broader deployment and potential commercialization, shown in Table 4-13.

Table 4-13. *DPMS Metric for Churn Prediction API*

Metric	Weight	Score	Weighted Contribution
Usage: 80% of customer success teams use it weekly	18%	0.80	0.14
Business value: Estimated €500K retention lift per quarter	25%	0.85	0.21
Reliability: 99.2% uptime, < 1% API error rate	17%	0.99	0.17
Maintenance: 6 hrs/month	12%	0.90	0.10
Adoption: Used in 3 regions, 2 additional business units onboarding	13%	0.70	0.09
Governance: Controls, lineage, and explainability in place	15%	0.80	0.12

DPMS = 0.83. A score of 0.83 places the churn model firmly in the mature data product category.

This means:

- Reliability and usage are strong
- Business value is proven and measurable
- Adoption is expanding

- Maintenance is predictable
- The product is capable of scaling
- Commercialization or broader internal rollout is warranted

This maturity assessment becomes the foundation for investment decisions and prioritization of next steps—whether that means extending the feature set, externalizing as a paid API, or integrating it deeper into enterprise workflows.

Why DPMS Matters Going Forward

As the organization scales its data product portfolio, DPMS plays three critical roles:

- **It creates accountability:** Teams can no longer call something a "product" unless it meets defined maturity criteria.
- **It drives investment discipline:** High-scoring products attract funding; low-scoring ones face redesign or retirement.
- **It operationalizes product thinking:** Engineering, product, compliance, and go-to-market teams share a common language of maturity and readiness.

DPMS becomes a strategic instrument guiding which data products are ready to commercialize, which require hardening, and which should be sunset. It replaces intuition with measurable direction and ensures that the organization evolves toward a disciplined, scalable, and value-driven data product ecosystem.

Conclusion: From Insights to Strategic Assets

By reframing data as a product ecosystem rather than a collection of analytical outputs, Richard's organization unlocked a fundamentally different way of creating value. Data products were no longer dashboards to be maintained; they were strategic assets capable of driving revenue, shaping customer experiences, and differentiating the business in the market.

The journey required a shift in mindset and operating model:

- Monetization became a structured business discipline rather than a hope.
- Economics and pricing were grounded in cost models, market sizing, and customer value.
- Operationalization and governance ensured reliability, trust, and scalability.
- Maturity assessment through DPMS provided a standard language for evaluating readiness and prioritizing investments.
- Product thinking replaced project thinking, creating a sustainable foundation for innovation.

With these foundations in place, the organization had established the essential building blocks of a modern data-driven enterprise—one capable of generating intelligence at scale, delivering curated insights, and bringing monetizable value to internal and external customers.

But data products were only the beginning.

As the landscape evolved, Richard recognized a deeper shift underway: intelligence itself was becoming productized, and AI was no longer an experimental capability—it was emerging as a fundamental driver of competitive advantage.

If data products taught the company how to build, monetize, and scale insights, then AI would challenge it to rethink how intelligence is created, delivered, and embedded into customer experiences.

The next chapter explores that frontier. The intersection of data, AI, and product design is where the real transformation begins.

Note If you can't measure a data product's maturity, you're not managing a portfolio, you're babysitting experiments.

CHAPTER 5

Beyond the Buzz: The Real Intersection of Data, AI, and Products

After several months of disciplined, iterative product development—with cultural battles fought, mindsets shifted, and teams reorganized into autonomous, empowered squads—Richard's analytics organization had finally become what he'd envisioned: a *product-led* team.

The transformation wasn't just technical, it was cultural. Stakeholders no longer thought of analytics as "reports on request." Business leaders now recognized data as a strategic asset, and every domain had a dedicated data product team that owned their products end-to-end. Roadmaps were aligned to business outcomes. Technical debt was down. Adoption was up. For the first time in years, the company's analytics capability felt scalable, predictable, and valuable.

He wanted to bring data-centric AI principles into every product: embedding machine learning models into the platform, refining feature engineering pipelines, and preparing for a future where AI wasn't a project, but a native capability baked into the company's DNA.

S. Shivam, *The Leadership Algorithm*, https://doi.org/10.1007/979-8-8688-2897-3_5

But just as his team was getting ready to make that pivot, ChatGPT happened!

Within weeks of OpenAI's groundbreaking release, the executive conversations shifted dramatically. Leaders who'd barely understand data pipelines were suddenly talking about transforming entire business units with AI. Boardrooms buzzed with plans for AI-driven marketing, AI-powered sales assistants, AI-generated code, and AI chatbots for customers.

Richard watched the hype curve rise with fascination—and a fair amount of dread.

Suddenly, AI wasn't a quiet enabler anymore; it was a board-level obsession. Senior leaders were sending him articles, case studies, and consulting reports about generative AI transforming industries overnight. Vendors pitched tools that claimed to "replace 70% of your workforce." Colleagues confidently declared that "AI will handle all forecasting from now on"—even though none of them could explain how the company's existing data pipelines worked.

Richard called it "**the leadership paradox**."

Ironically, the timing couldn't have been worse. Richard's team had just reached a level of data maturity that positioned them to operationalize AI responsibly. They had:

- Clean, governed datasets for critical business domains
- A portfolio of successful machine learning products (fraud detection, churn prediction, forecasting)
- Automation pipelines that made experimentation easy
- A strong product culture that kept analytics aligned to business outcomes

But those accomplishments were invisible in the shadow of generative AI hype. Executives weren't asking about fraud detection or churn prevention anymore—they wanted ChatGPT clones and AI copilots.

For Richard, this leadership journey wasn't just about embracing the new wave of technology. It was about re-educating an organization that had forgotten what AI truly was.

Richard realized the company had walked straight into the global problem:

AI had become synonymous with GenAI, and in the process, leaders were losing sight of what AI actually is, what forms it takes, and where the real value lies.

The hype drowned out the fundamentals.

To reset the conversation, Richard stood up, walked to the whiteboard, and wrote a single sentence:

AI is simply the ability for machines to learn patterns and make decisions. Not magic. Not science fiction. Not a chatbot that happens to write emails.

AI, at its core, is a spectrum of capabilities that help businesses:

- Predict
- Classify
- Rank
- Recommend
- Optimize
- Generate
- Understand

The problem wasn't that leaders didn't believe in AI.

It's that they didn't understand it well enough to *use* it effectively.

He needed to show them:

- That AI wasn't new; many of their highest-value products were already powered by ML.
- That generative AI was just one layer of a larger intelligence stack.
- That without data-first discipline, all these grand visions would collapse under their own weight.

"Look at this demo," the COO said, spinning his laptop around to face the group. On the screen, a chatbot interface generated a marketing plan in seconds. "This is incredible. Why aren't we using this already?"

The CMO leaned forward. "We could use AI to auto-generate all our ad copy. Imagine the cost savings. Entire campaigns, done in minutes!"

The CTO, not to be outdone, chimed in, "There's a startup claiming they can train a model to rewrite our legacy codebase. We should explore that."

Then came the CEO's declaration, "Every business unit should submit a plan by next month to integrate AI into their workflows. Let's be an AI-first company."

Heads nodded. Excitement buzzed around the table. Everyone was already imagining efficiency gains, automation, and headlines about their company being a leader in AI.

This wasn't the first time Richard had seen an industry frenzy. He'd been through the "Big Data" gold rush, the Hadoop hype cycle, and the blockchain experiments that led nowhere. But this time, the confidence in AI was almost religious. Executives who couldn't explain how their dashboards worked now spoke about neural networks with certainty.

Richard stayed quiet; he knew that behind every AI success story was an unglamorous truth: clean, connected data, rigorous governance, engineering discipline, and years of trial and error. None of those realities were reflected in the conversation unfolding before him.

Richard knew better than to fight the energy in the room. The sudden wave of enthusiasm for AI wasn't a problem in itself—it was a gift. **For the first time in years, the entire leadership team cared deeply about analytics and technology**. But left unchecked, this enthusiasm would create scattershot projects: expensive experiments, redundant tooling, and proof-of-concepts that would never see production.

If there was ever a moment to architect an enterprise AI strategy, this was it. And so began the next phase of his mission. Before discussing strategies, architectures, or product implications, Richard realized the organization needed something simpler: a shared, realistic understanding of what AI is and what it is not.

AI is not a single technology. It is a collection of techniques that help machines identify patterns, make predictions, automate decisions, and generate new content. GenAI is one part of this ecosystem—not the ecosystem itself.

To reestablish a shared vocabulary, Richard started using a simplified framework in discussions with leadership teams. It avoided technical depth and placed each form of AI in terms of practical business outcomes.

The framework had four pillars, discussed in the following sections.

Pillar 1: Predictive AI (Supervised Machine Learning)

This is the form of AI that is most responsible for measurable business value. Its purpose is straightforward: predict an outcome. Examples include:

- Lead conversion probability
- Employee attrition likelihood
- Credit risk
- Fraud scoring
- Demand forecasting

These models do not generate content. They predict what is likely to happen next—which is often exactly what business decision-making requires.

For many leaders, this was a useful reminder: the AI behind sales funnels, underwriting, supply chain decisions, and customer retention is not generative at all.

Pillar 2: Pattern Discovery (Unsupervised Learning)

This branch of AI identifies structure where there are no predefined labels. Its value lies in:

- Customer segmentation
- Behavioral grouping
- Anomaly detection
- Market and cohort analysis

These capabilities frequently power personalization, product recommendations, risk segmentation, and fraud monitoring pipelines.

Pillar 3: Optimization and Automation (Decision AI)

Here, AI is used to automate or optimize decision-making within a defined set of constraints. Examples include:

- Inventory optimization
- Routing and scheduling
- Automated credit approvals
- Repricing engines

This is the AI that runs quietly in the background of operational systems.

Pillar 4: Generative AI (LLMs, Diffusion Models)

This pillar has captured global attention for its ability to generate:

- Text
- Summaries
- Code
- Images
- Synthetic data

It excels at tasks involving communication, synthesis, content creation, and reasoning through natural language.

When product leaders cannot differentiate these pillars, three things happen:

- They apply GenAI to the wrong problems.
- They overlook predictive models that directly improve KPIs.
- They fail to see how AI integrates into workflows beyond chat interfaces.

For example:

- Lead management relies on predictive scoring, not text generation.
- Attrition prediction comes from classification models, not LLMs.

- Fraud monitoring depends on anomaly detection and supervised learning.
- Recommendations come from ranking models, not content generators.

Two Practical Examples: Predictive AI vs. Generative AI in Real Enterprises

As Richard continued working with business units, he noticed that abstract explanations of AI weren't enough. Leaders needed concrete, practical examples that reflected problems they handled every day.

Two use cases helped clarify the distinction better than any slide deck:

- Customer retention prediction for a large retail enterprise
- AI agent for digital customer support in a banking environment

Both carried the "AI" label, but the underlying technology, data, and business impact came from completely different branches of AI.

Example 1: Customer Retention Prediction for a Retail Giant

In retail, customer churn is a constant concern. Richard often described a scenario that resonated:

A national retail chain wants to predict which customers are likely to stop shopping with them in the next 60–90 days.

This is not a conversational problem. It is not a summarization problem. And it cannot be solved by asking as LLM a clever prompt. It is a predictive machine learning use case.

How the Problem Is Solved (Conceptually)

A predictive model uses structured data such as:

- Loyalty card activity
- Purchase frequency
- Basket size
- Recency of transactions
- Product categories browsed vs. purchased
- Coupon redemptions
- Store visit frequency
- Online vs. offline behavior
- Customer service interactions
- Local competitor density
- Price sensitivity patterns

The model learns patterns from customers who churned in the past and applies those patterns to current customers.

The output is numerical:

- "Customer 33149 has a 72% chance of leaving in the next 30 days."
- "Customer 10412 shows improving retention likelihood."

The model doesn't write anything. It doesn't explain itself in natural language. It simply predicts probability.

Where the Value Comes From

The business value in retail churn prediction comes from:

- Identifying at-risk customers
- Targeting them with offers or loyalty programs
- Preventing revenue leakage
- Improving marketing spend efficiency
- Prioritizing retention campaigns

Example 2: AI Agent for Digital Customer Support in Banking

Richard paired the retail example with a second one from financial services, one that clearly illustrated where LLM-based agents' shine.

A large bank wants to deploy a digital support agent capable of handling customer inquiries through chat, mobile apps, or voice assistants. Unlike the retail churn model, this problem is not about prediction. It is about:

- Understanding language
- Retrieving relevant information
- Holding multi-step conversations
- Completing tasks on behalf of a customer

This is agentic AI territory—powered by LLMs, retrieval systems, and workflow automation.

How the Problem Is Solved (Conceptually)

An AI banking support agent typically combines:

1. An LLM to interpret customer questions:
    - "Why was my card declined?"
    - "How can I increase my transfer limit?"
    - "Show me my last three transactions."
2. A Retrieval System (RAG) to pull policy documents, FAQs, and customer-specific information:
    - Card policy
    - Fraud rules
    - Minimum balance requirements
    - Loan terms
    - Spending history (via secure internal APIs)
3. Action execution capabilities to perform tasks:
    - Freeze card
    - Reset passwords
    - Initiate transfers
    - Update personal details
    - Raise support tickets
4. Guardrails and compliance logic critical in banking:
    - KYC
    - Fraud checks
    - Authentication
    - Privacy constraints
    - Regulatory boundaries
    - Escalation rules

This is an agent—not a predictor.

Where the Value Comes From

The business value in banking support agents includes:

- Reduced call center load
- Faster resolution times
- Better customer experience
- 24/7 availability
- Consistent responses
- Lower operating costs
- Improved self-service adoption

Underlying these examples are three critical insights, covered in the following sections.

Insight 1: Predictive ML and GenAI Solve Different Classes of Problems

- Retail churn prediction needs structured data and statistical learning.
- Banking support agents need language models, retrieval, and workflows.

Insight 2: Predictive AI Drives Business Metrics; GenAI Drives Workflows

- Predictive AI improves KPIs like revenue, churn, fraud, and efficiency.
- Generative AI improves customer experience, operations, and productivity.

Insight 3: The Future of AI Products Combines Both

- Predictive ML identifies the what.
- Generative AI (LLM) assists with the how.

Imagine the retail example enhanced with an agent:

- Predictive AI identifies customers at risk.
- AI agent drafts personalized retention campaigns.
- Agent executes communications across channels.

Or in banking:

- Predictive AI flags potential fraud.
- AI agent interacts with the user, verifies details, and escalates if needed.

This complementary relationship is where the next wave of AI product innovation lies.

By the time Richard had aligned his leadership teams on the fundamentals of AI, a more practical question emerged—"Where does AI actually move the needle for the business?"

This wasn't about hype or theoretical capability. Executives wanted to understand where AI could deliver real, defensible, measurable business impact—not interesting prototypes or "innovation theatre."

Richard framed the answer around three value zones that consistently show up across industries. These zones helped product and business leaders understand not just what AI can do, but when to use it and how to evaluate opportunities.

Value Zone 1: Efficiency: Doing the Same Work, But Faster and Cheaper

This is the most straightforward—and often the fastest—source of AI value. AI improves efficiency through:

- Automation
- Streamlined workflows
- Smarter routing and triage
- Document understanding
- Reducing manual or repetitive work
- Faster decisions based on existing rules

Table 5-1 lists some of the examples that resonated internally.

Table 5-1. *Value Zone 1 Examples*

Type	Examples
Customer operations	AI-assisted support agents Automated email categorization Claims document extraction Fraud alert triaging
Internal productivity	Generating reports Summarizing meetings Drafting internal communication Preparing compliance documentation
Back-office processes	Automated reconciliation Invoice classification Contract analysis

Richard often reminded the team that efficiency gains are not glamorous, but they are reliable. These initiatives don't win awards, but they consistently deliver measurable savings. In many organizations, this is the first step in their AI journey—reshaping everyday operations.

Value Zone 2: Effectiveness: Better Decisions, Better Outcomes

This is where predictive AI dominates. The goal is not speed or cost reduction, but better decisions than those driven by gut instinct or rigid business rules. Examples across industries include the ones outlined in Table 5-2.

Table 5-2. *Value Zone 2 Examples*

Type	Examples
Retail	Inventory forecasting Dynamic pricing and markdown optimization Customer churn prediction Purchase propensity scoring Store labor planning Assortment optimization
Banking and financial services	Credit risk scoring Fraud detection Claims routing Customer lifetime value estimation Loan approval ranking Collections prioritization

(*continued*)

Table 5-2. *(continued)*

Type	Examples
Telecom	Network optimization and congestion prediction Intelligent tower load distribution Churn propensity models Product bundle recommendations Proactive outage prediction Cross-sell and upsell scoring
Insurance	Underwriting risk scoring Claims fraud likelihood Premium pricing optimization Policy renewal prediction Loss ratio forecasting Exposure modelling Severity prediction for auto, health, property claims
E-commerce	Personalized product ranking Search relevance optimization Cart abandonment prediction Inventory availability forecasting Real-time pricing Return likelihood prediction Fraud scoring for payments Recommendation systems based on browsing and purchase patterns

(continued)

Table 5-2. *(continued)*

Type	Examples
Dining and quick-service restaurants	Daily and hourly demand forecasting Inventory and food preparation planning Menu item recommendation based on weather, time, and location Dynamic staffing optimization Delivery time prediction Customer loyalty scoring Meal personalization models

Effectiveness-driven AI improves KPIs that senior leaders care deeply about:

- Revenue
- Margin
- Retention
- Loss ratios
- Conversion
- Utilization

Richard emphasized something product leaders often underappreciate: The biggest AI wins don't come from what the model does. They come from what the business does differently because of it. Better interventions lead to better outcomes.

Value Zone 3: Innovation: New Capabilities, New Products, New Business Models

The third zone is where AI stops supporting the business and starts redefining it. These are the opportunities that:

- Create new revenue streams
- Differentiate products in the market
- Shift customer expectations
- Unlock new business models

Examples are listed in Table 5-3.

Table 5-3. *Value Zone 3 Examples*

Type	Examples
Industry-specific intelligence products	Fraud-as-a-service Inventory optimization engines Credit decisioning modules Benchmarking and predictive analytics bundles
AI-powered product features	Personalized experiences (retail, fitness, edtech) Conversational interfaces (banking, telecom) Automated financial coaching Intelligent document assistants for legal and insurance
Entirely new product categories	AI copilots for knowledge workers Vertical-specific AI agents Smart personalization engines Multimodal experience layers (voice, text, and images)

This is where both predictive AI and generative AI (LLMs) converge to build new value, not just optimize existing workflows. Richard often described this zone as "AI enhanced product strategy, not AI pilots."

To simplify, Richard used a simple test, which he called a practical rule for product leaders, outlined in Table 5-4.

Table 5-4. *Decision Table for Predictive ML vs. LLM vs. Agentic AI*

Problem	Method/ Approach	Example
If the problem is numerical	Predictive ML	Conversion rates, churn, fraud, pricing, forecasting
If the problem is linguistic or reasoning-based	LLMs	Support queries, summarization, document understanding
If the problem requires multi-step action	Agentic AI	Onboarding workflows, customer service flows, and compliance tasks.
If the problem is about structure	Unsupervised learning	Segmentation, clustering, anomaly detection

This simple classification helped product teams avoid misuse, overuse, and misdirection and move toward AI opportunities grounded in business value.

AI As a Product Capability, Not As a Lab Experiment

As interest in AI grew across the organization, Richard noticed something common and entirely understandable: most teams were enthusiastic, but unsure how AI would fit into the products they were responsible for.

To many, AI still felt like something separate—a specialized activity belonging to data scientists or innovation groups. Team members often asked whether AI needed a dedicated project stream, a different process, or a separate roadmap. Others wondered if AI meant building entirely new products rather than enhancing existing ones.

Richard recognized this as a natural stage of adoption. When a capability is new, people want to understand where it belongs and how to work with it. He approached the topic the same way he did when introducing any unfamiliar discipline—by simplifying it and making it relatable to how product teams already think.

The first idea he emphasized was this: AI is not a standalone initiative. It is a product capability, similar to search, personalization, scoring, analytics, or automation.

We don't treat search as a separate "search project"; we treat it as a feature set embedded into product experiences. We don't isolate personalization as a lab experiment; it's part of the user journey. AI follows the same pattern. It becomes valuable not because it exists, but because it elevates something the user is trying to accomplish.

To help teams frame this correctly, Richard encouraged them to start with the product experience rather than the model. Instead of asking, "What AI can we build?" he encouraged the question, "What part of the user journey becomes better, faster, or more intelligent with AI?" This reframed conversations in ways that designers, engineers, and product managers could immediately relate to. A predictive score is useful only when it appears at a step in the workflow where someone needs to make a decision. A summary is valuable only when it reduces cognitive load in a process that involves too much reading or context switching. An AI-generated recommendation becomes meaningful only if it aligns with user intent and is surfaced in the right moment. The capability had to fit naturally into existing product reasoning.

Richard also explained that AI behaves more like a system than a one-time deliverable. Unlike traditional features that remain stable once built, AI models evolve with data, behavior, and context. Predictive models need occasional retraining. Agentic workflows require refinement as new user scenarios emerge. LLM prompts, guardrails, and retrieval strategies improve over time. Because of this, he encouraged teams to treat AI capability the way they treated performance optimization or ranking logic—as something that matures through iteration rather than something that is "*completed*."

To make this easier organizationally, Richard advocated for a hybrid structure: a central AI platform team responsible for shared tooling, data pipelines, governance, security, and operating standards; and product teams who embed AI capabilities into their roadmaps where appropriate. This allowed teams to move at their own pace while maintaining consistency in how AI was built, deployed, and monitored. It also helped product managers build confidence by giving them clear ownership without expecting them to become machine learning experts (see Table 5-5).

Table 5-5. *What the AI Platform Team Owns vs. What Product Teams Owns*

Capability Area	AI Platform Team Owns	Product Teams Own
Core AI infrastructure	Shared compute environment, model hosting platforms, vector databases, orchestration layers, secure runtime environments	Consumption of approved platform services within their products
Model registry and lifecycle	Central model registry, approved foundation models, version control, deployment standards, rollback mechanisms	Selection of appropriate approved models for product use cases, domain tuning decisions

(*continued*)

Table 5-5. *(continued)*

Capability Area	AI Platform Team Owns	Product Teams Own
Agent/tool registry	Registry of reusable agents, tools, connectors, MCP interfaces, shared skills library	Composition of agents into domain workflows and user experiences
Feature stores/ data foundations	Shared feature stores, reusable training data assets, governed data pipelines, common semantic layers	Domain-specific features, business rules, proprietary context signals
Evaluation frameworks	Standard evaluation harnesses, benchmark suites, quality scorecards, regression testing pipelines	Use-case success metrics, business outcome measurement, acceptance criteria
Guardrails and responsible AI	Safety controls, policy enforcement, prompt filtering, PII controls, access governance, explainability standards	Applying controls correctly within product flows and escalating exceptions
Observability and monitoring	Central monitoring for latency, cost, drift, hallucination risk, uptime, agent behavior, logging standards	Monitoring user impact, workflow success rates, product adoption signals
Product prioritization	Enterprise capability roadmap and reusable platform priorities	Business use-case prioritization, ROI cases, roadmap sequencing
Business logic	Generic orchestration patterns only	Domain workflows, rules, approvals, exception handling, decision logic

(continued)

Table 5-5. *(continued)*

Capability Area	AI Platform Team Owns	Product Teams Own
User research and experience	Shared design patterns for AI experiences	Deep customer research, journey mapping, UX design, trust signals, human handoff moments
Change management and adoption	Training playbooks, internal standards, enablement materials	Adoption within business unit, stakeholder engagement, frontline rollout
Value realization	Platform efficiency, reuse, scalability, governance maturity	Revenue impact, productivity gains, customer experience improvement

Another topic that Richard frequently clarified was the importance of defining the role of AI within a user experience. Teams sometimes imagined AI as a "smart layer" that would transform the entire product, when in reality, most successful AI adoption comes from small, well-placed enhancements. A fraud score inside a decision screen. A generated summary inside a service console. A ranking model that decides the order of displayed items. These are not dramatic or visible, yet they dramatically improve workflow efficiency and decision quality. Richard encouraged teams to aim for "subtle intelligence" rather than "grand AI moments."

To support better decision-making, he offered a straightforward reflective guide. Before adding AI to a product, teams should ask:

- Where exactly will the AI output appear?
- Who will use it and at what moment?

- What decision or task becomes easier because of it?
- How will we know if the AI is helping?
- What needs to be monitored or updated over time?

These questions did not require any expertise in machine learning—only clarity about product behavior and user needs. They helped teams avoid overengineering and focus on the parts of the experience where intelligence actually matters.

Over time, this teaching-oriented approach helped teams see AI the same way Richard did: not as a separate discipline, but as an ingredient that blends into the broader product experience. The goal was not to build "AI features," but to build thoughtful, useful experiences that happened to use AI behind the scenes. When AI was understood as a capability—adaptable, incremental, and closely tied to user workflows—teams were able to integrate it confidently and responsibly.

This foundation prepared them for the next challenge: Understanding how to design product experiences that incorporate intelligence in ways that feel natural, trustworthy, and genuinely helpful to users.

Designing Intelligence Inside Products

As teams began to understand AI not as something separate but as a capability that blended into their products, the next logical question emerged, "How should intelligence actually appear inside a product experience?" AI can influence design in many ways—subtle, visible, interactive, or fully behind-the-scenes—and product leaders needed clarity on how to integrate it thoughtfully.

Richard approached this as a design conversation rather than a technical one. Many teams initially imagined AI as a prominent feature: a chatbot icon, a "smart" button, or a dedicated panel that did something intelligent.

But most mature AI products don't look like AI products. They look like ordinary tools that happen to feel more responsive, helpful, or tailored. The intelligence is present, but not always visible.

He encouraged teams to begin with a simple principle: Intelligence should appear where users naturally need support, not where the technology is easiest to showcase.

That meant understanding the workflow, not just the model. When users struggle to interpret information, AI might summarize or highlight key insights. When they need to make decisions, AI might provide predictions or recommendations. When tasks involve too many steps or repetitive actions, AI might automate or pre-fill portions. The design starts with the friction, not the algorithm.

Richard often explained that there are roughly four ways AI can show up inside a product—each serving different purposes and requiring different design considerations. The following sections cover these four approaches to intelligence.

Intelligence As a Background Capability

Some AI capabilities operate almost invisibly. Ranking engines, fraud scoring, search relevance, and personalization models often fall into this category. Users don't see the model; they see the outcome. For example:

- A search bar returning more relevant results
- A homepage showing personalized recommendations
- A decision workflow automatically highlighting items of concern
- A routing engine quietly assigning tickets more effectively

In these cases, the intelligence doesn't need to be explained extensively unless required for trust or compliance. The value lies in a smoother experience, not a more visible model.

Richard advised teams to treat these capabilities like electricity—essential, influential, but not the focus of user attention.

Intelligence As a Decision Support

In other cases, AI provides information that assists the user in making a better decision. This includes predictive scores, alerts, risks, recommendations, and forecasts. Here, the way AI is presented matters. When showing a prediction, users need context:

- What does the score mean?
- Why is it useful right now?
- How confident should they be in it?
- What action does it support?

Richard encouraged teams to embed predictions directly into the decision points where humans already consider information. Instead of a separate dashboard with "AI insights," the output should appear inside the normal workflow—the screen where someone approves a loan, reviews a customer, prioritizes a task list, or evaluates a case. He reminded teams that a prediction alone is not a solution; it becomes meaningful only when paired with the next step a user can take.

Intelligence As a Copilot (Interactive Assistance)

With the rise of LLMs, AI can now help users through interactive assistance—summarizing, drafting, generating options, interpreting documents, or walking users through multi-step tasks. This is where "AI copilot" functionality often emerges. Typical examples include:

- Drafting customer messages
- Summarizing long documents
- Translating policy into plain language
- Preparing call notes or follow-up actions
- Helping a user find information across multiple systems

Here, design choices shape trust and adoption. Users need clarity on:

- What the AI can do
- What it cannot do
- How to edit or override its output
- Whether the AI is retrieving real company information or generating based on general knowledge

Richard advised teams to frame the AI as a helper, not an authority. The tone should suggest collaboration ("Here are three options you can refine.") rather than certainty ("This is the correct answer."). The goal is to boost productivity without implying that the system fully replaces human judgment.

Intelligence As an Agentic AI (Task Automation)

The most advanced form of AI integration is an agent that can complete tasks on behalf of the user: opening tickets, updating records, performing steps across systems, or orchestrating actions. These agentic capabilities blend reasoning (LLMs) with structured execution logic. Examples include:

- Resetting customer passwords
- Updating account details

- Creating case notes
- Triaging issues
- Initiating workflows or approvals

Designing AI agents introduces additional considerations:

- Transparency: What did the agent do?
- Reviewability: Can the user undo or inspect steps?
- Control: When does the agent act autonomously vs. with confirmation?
- Boundaries: Which tasks require manual approval?

Richard often summarized agent design with a simple guideline: An agent should reduce effort, not remove visibility. Users need to feel supported, not sidelined.

The Economics of AI Products: Pricing to Value

While designing intelligence into products is an important step, many organizations struggle with an equally critical issue—how to quantify the value of AI and translate it into a pricing strategy that makes sense for customers and for the business.

AI is often discussed in terms of innovation and capability, but far less attention is given to understanding how different AI components drive cost, how value should be measured, and how pricing models can be used strategically to differentiate products in the market.

Richard noticed that even experienced product teams were not always comfortable navigating this territory. Traditional SaaS economics—where marginal cost approaches zero—don't map cleanly to AI-driven experiences. AI introduces variable compute costs, evolving data needs, and ongoing refinement cycles that require a different economic lens.

The goal wasn't to turn product managers into financial analysts or ML engineers, but to help them develop a practical understanding of how AI affects both sides of the equation—the cost to deliver the capability and the value it creates for the customer.

Why AI Requires a Different Economic Frame

AI systems behave differently from traditional software, and so do their economics. Predictive models, LLMs, and agents all incur costs not only to build, but also to operate. There is a *real*, ongoing cost every time intelligence is used—predictions computed, tokens processed, embeddings generated, or documents summarized.

There are also long-term costs tied to:

- Data acquisition and storage
- Monitoring
- Retraining
- Governance
- Labeling and feedback loops
- Continuous improvement

Because of this, AI products cannot be priced as one-time features or as generic software modules. Their economics depend on scale, usage patterns, and the business value they enable.

Understanding this helps teams avoid two common traps:

- Underpricing AI features because they assume marginal cost is negligible.
- Overpricing them because they assume AI is inherently premium.

The right price emerges from understanding cost drivers and matching them to value delivered.

The Four Drivers of AI Cost

Richard explained AI cost structure through four core drivers that apply consistently across industries, covered next.

Compute Cost (Inference Cost)

Every prediction, generation, or workflow execution triggers compute usage.

- Classical ML: Low per-call cost
- Small LLM: Moderate cost
- Large LLM: Higher cost
- Agentic workflows: Multiple calls, orchestration, retrieval, and tool usage

This is often the most variable and misunderstood cost component.

Data Infrastructure Cost

AI capabilities rely on a healthy data foundation. Costs accumulate in:

- Cloud storage
- Pipelines
- Feature stores
- Vector databases
- Semantic layers
- Data catalogs and lineage systems

These costs grow with data volume, product breadth, and increasingly with new data generated by conversational products, interaction logs, memory layers, and synthetic data created for training, testing, or tuning models and agents.

Operations, Maintenance, and Monitoring Cost

AI is dynamic and requires continuous operational investment. Costs include:

- Drift monitoring
- Model retraining
- LLM prompt optimization
- Guardrail updates
- Periodic evaluations
- Incident responses

This produces recurring operational spend. In mature organizations, this also expands into MLOps/LLMOps/AgentOps capabilities such as:

- CI/CD pipelines for models and prompts
- A/B testing infrastructure
- Evaluation datasets and benchmark harnesses
- Rollback and failover mechanisms
- Observability tooling for quality, latency, and cost

These recurring costs often exceed raw inference spend at scale.

Human Expertise Cost

Behind every AI capability sit teams that ensure that it remains robust and responsible:

- ML engineers
- Data scientists
- Product managers

- Domain experts
- Responsible AI reviewers
- Prompt/agent designers

These costs often determine the sustainability of an AI product.

Understanding these levers helps product leaders model margins before deciding how to package or charge for an AI capability.

Pricing Models That Work for AI Products

AI pricing is not universal; it must align with usage, cost patterns, and customer perception of value. Richard described the following four models that most enterprises adopt.

Model 1: Per-User Pricing (AI Assistants and Copilots)

Suitable for LLM-enhanced workflows that improve individual productivity.

Example: $20–$40 per user per month for AI-assisted workspace tools.

Model 2: Usage-Based Pricing (API-Driven AI)

Best for predictive scoring, document extraction, retrieval, embeddings, or agentic calls.

Example:

- $0.001 per predictive score
- $0.02 per LLM call
- $0.15 per automated agent workflow

This aligns cost with usage and gives customers transparency.

Model 3: Tiered Bundles (Platform Products)

Good for mixed workloads where users want predictable billing.

Examples:

- Basic: Limited AI calls
- Professional: Full predictive and moderate LLM
- Enterprise: Custom models and automation

Model 4: Outcome-Based Pricing (High Impact AI)

Used when AI connects clearly to a financial outcome.

Examples:

- Retention uplift priced as a percentage of revenue saved
- Fraud AI priced as share of fraud prevented
- Dynamic pricing priced from margin uplift

This model is powerful but requires strong measurement discipline.

The "Pricing to Value" Ladder

Richard often helped product leaders visualize AI value and pricing power through a simple ladder.

As intelligence moves up this spectrum:

- The cost to deliver increases
- The complexity increases
- The risk increases
- The value to the customer increases even faster

Pricing should follow the progression shown in Figure 5-1.

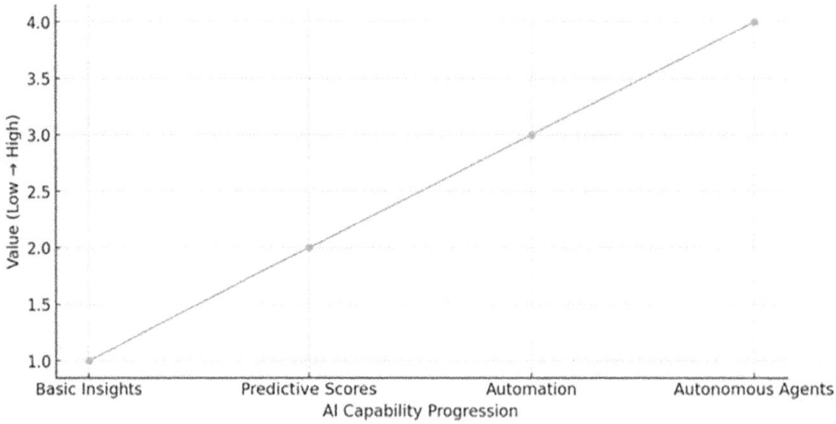

Figure 5-1. *Pricing to AI value ladder*

At the end of these discussions, Richard always emphasized one principle: Customers do not pay for models; they pay for outcomes.

They pay for:

- Fewer fraud losses
- Higher retention
- Smoother customer support
- More efficient operations
- Personalized experiences
- Automated decisions

The AI behind the scenes—whether predictive, generative, or agentic—is simply the engine that delivers that outcome. Understanding this frees teams from obsessing over model cost alone and helps them focus on pricing relative to business value delivered.

Richard often emphasized that pricing discussions become meaningful only when tied to ROI—not feature lists, not model complexity, not "AI premium," but the actual business value created. Without ROI, pricing is guesswork; with ROI, it becomes a strategic decision.

AI ROI Is Dynamic, Not Static

AI ROI differs from traditional software ROI because:

- The value depends on behavioral change (better decisions, faster workflows).
- The impact may be distributed across multiple teams.
- Costs scale with usage.
- Benefits can compound over time (e.g., incremental retention).
- There is often a measurable financial outcome (e.g., saved losses, reduced handle time).
- ROI can also decay over time if it's not actively managed.

Richard reminded the teams that strong ROI in quarter one does not guarantee strong ROI in quarter four. Models drift, customers become less responsive, competitors catch up, and workflows evolve. Common causes of ROI decay include:

- Model drift reducing prediction quality
- Customer fatigue from repetitive recommendations or messaging
- Changing market conditions or seasonality

- Competitors matching your capability
- Poor adoption after initial launch excitement
- Rising inference or servicing costs

For this reason, AI products require continuous refresh, experimentation, and optimization rather than one-time deployment.

Richard's Three ROI Axes

Richard taught teams to evaluate ROI across three axes:

1. Direct Impact
    - Revenue generated
    - Cost avoided
    - Margin uplift
    - Productivity gains
2. Operational Impact
    - Faster decisions
    - Reduced manual effort
    - Lower error rates
    - Improved throughput
3. Experience Impact
    - Better customer experience
    - Higher satisfaction
    - Reduced friction
    - More consistent outcomes

Pricing should correlate with the strength of ROI along these axes and the organization's ability to sustain that ROI over time. To make this practical, Richard walked teams through the following concrete ROI models for three very different AI product types.

ROI Example 1: Customer Attrition Prediction (Insurance/Predictive ML)

Attrition is one of the most expensive problems in insurance. Acquiring each new policyholder can cost five to seven times more than retaining an existing one. Predictive ML helps identify at-risk customers early—but how do you quantify ROI?

Richard broke it down this way.

Step 1: Define the Predictive Output

A churn model provides a probability score:

- "Customer 10342 has a 72% chance of attrition within 90 days."

Step 2: Map the Score to a Real Intervention

Typical interventions:

- Personalized outreach
- Agent call with tailored offers
- Targeted discount
- Policy review session
- Loyalty incentive

These interventions have a cost.

Example:

- Outreach campaign: €3 per customer
- Agent call: €10 per call
- Retention offer: €25 per saved customer

Step 3: Estimate Retention Impact

Suppose:

- 100,000 policyholders
- 10% annual attrition (10,000 customers)
- Churn model identifies 5,000 high-risk customers
- Intervention saves 15% of them (750 customers)

Step 4: Estimate Financial Impact

If average annual premium is €1,000, then saving 750 customers leads to €750,000 retained revenue.

Subtract intervention cost:

- 5,000 outreach actions × €3 = €15,000
- 1,000 escalations × €10 = €10,000
- Retention incentives for 750 customers × €25 = €18,750

The total intervention cost is €43,750. The net ROI is ≈ €706,250.

Step 5: Determine AI Pricing Boundaries

Since the customer earns more than €700k in net retained revenue, one of the following could happen:

- AI provider could charge €50k–€150k annually
- €1 per scored policyholder
- Outcome-based pricing: 5–10% of revenue retained (€35k–€70k)

Anything that keeps cost far below generated value is acceptable.

ROI Example 2: Customer Support AI Agent (Banking/GenAI and Agentic Workflow)

Customer support is one of the highest cost functions in financial services. AI agents dramatically reduce load by handling routine or moderately complex queries. But the economics differ from predictive ML.

Step 1: Define What the AI Agent Can Do

Capabilities:

- Interpret customer questions
- Retrieve information
- Summarize documents
- Complete simple tasks (e.g., blocking card, updating address)
- Escalate complex cases with notes

Step 2: Establish Baseline Costs

Assume:

- Avg cost per human-handled support call is €4.
- Annual call volume is 5 million calls.
- 60% are routine, so 3 million calls are automatable.

Step 3: Estimate Agent Impact

AI agent handles:

- 40% of routine calls independently, which is 1.2 million calls.
- 20% assisted calls (AI drafts answer, agent reviews)—600,000 calls.

Impact:

- Fully automated calls: Cost drops from €4 = €0.20 per AI conversation.
- Assisted calls: Cost drops from €4 = €1.80.
- Remaining calls unchanged.

Step 4: Financial Impact

Fully automated calls:

- Savings per call: €4 - €0.20 = €3.80
- Total savings: 1.2M × €3.80 = €4.56M annual savings

Assisted calls:

- Savings per call: €4 - €1.80 = €2.20
- Total savings: 600k × €2.20 = €1.32M annual savings

Total ROI leads to €5.88M in operational savings per year.

Step 5: Cost of AI

Let's assume:

- 1.2M AI agent conversations @ €0.04 each = €48,000
- 600k assisted calls @ €0.02 each = €12,000
- Total LLM inference cost ≈ €60,000 annually
- Platform, monitoring, and integrations ≈ €150,000 annually

Total operational cost is ≈ €210,000.

Step 6: Real ROI

Savings: €5.88M
AI cost: €0.21M
Net ROI ≈ €5.67M
ROI multiple ≈ 27×

Step 7: Pricing Considerations

The bank would comfortably pay one of the following:

- €300k–€1M annually (subscription model)
- Usage-based pricing (e.g., €0.05–€0.10 per automated conversation)
- Outcome-based pricing tied to cost savings (e.g., 5% of savings = ~€294k)

Because the value created is so large.

ROI Example 3: Dynamic Pricing for Restaurants (QSR/Hospitality/Optimization and Predictive ML)

Restaurants, especially quick-service chains (QSRs), operate on tight margins, fluctuating demand, and intense competition. Dynamic pricing uses predictive models and optimization engines to adjust prices based on demand signals, time of day, footfall, weather, supply constraints, and local events.

Richard used this example to show how ROI appears in industries with high volume but low margin, where even small optimizations compound significantly.

Step 1: Define the AI Capability

Dynamic pricing combines:

- Demand prediction model: Forecasts order volume for each menu item per hour.
- Price elasticity model: Predicts how demand responds to price changes.
- Optimization model: Finds ideal prices that maximize revenue without hurting customer experience.

The AI suggests small (1–7%) price adjustments based on real-time conditions:

- High footfall: A slight price increase
- Low demand: Promotional pricing
- Inventory constraints: A shift in demand to substitutes
- Rainy days: An increase in hot beverage pricing
- Local events: Adjustments in popular item pricing

Step 2: Establish the Restaurant's Baseline

Consider a common scenario for a mid-sized restaurant chain:

- 80 restaurants
- Average daily revenue per store: €4,000
- Annual revenue per store: €1.46M
- Total chain revenue: €116.8M
- Typical profit margin: 6–8%

A 2–5% revenue uplift has a massive impact due to margin amplification.

Step 3: Estimate Impact of Dynamic Pricing

Let's assume:

- AI is applied to 30–40% of menu items (high-volume SKUs).
- Price adjustments range from ±1% to ±5%.
- Demand forecasting and elasticity modeling yield a 3% uplift across targeted items.
- Only 30% of total sales come from those targeted items.

Revenue impact calculation:

- Portion of revenue impacted: 30% of €116.8M = €35.04M.
- 3% uplift means €1.051M additional revenue per year.

Step 4: Estimate Profit Uplift

Because restaurant margins are tight, incremental revenue disproportionately improves profit. Assume:

- Profit margin without AI = 7%
- Profit on incremental revenue = ~70–80% (because fixed costs stay constant)

So, profit uplift ≈ €1.051M × 75% ≈ €788,000 net additional profit. This is *pure incremental margin.*

Step 5: Model the Cost of AI

Costs include:

1. **Platform subscription**

   Dynamic pricing engine: €120,000/year

2. **Per-restaurant integration and maintenance**

   Assume: €500 per location per year for 80 restaurants is €40,000.

3. **Compute (inference and optimization)**
   100k pricing calculations per day at €0.0005 each for ~€50/day is €18,250/year.
   Total AI cost: €178,250/year (rounded to €180k)

Step 6: Final ROI

- Additional profit ≈ €788k
- AI cost ≈ €180k

  Net ROI ≈ €608k

  ROI multiple ≈ 4.4× ROI payback period: < 3 months.

Step 7: Pricing Strategy for This Use Case

Given the strong ROI, an AI provider could use any of these options:

1. **Option 1: Per-location pricing**
    - €2,000–€5,000 per restaurant per year for 80 restaurants is €160k–€400k/year.

      Highly scalable and predictable.
2. **Option 2: Percentage-of-uplift pricing**
    - 10–20% of revenue uplift. If uplift = €1.051M then pricing is €105k–€210k.
    - Aligned with value creation.
3. **Option 3: Hybrid**
    - Base fee and small share of uplift (e.g., €100k base and 5% uplift).
4. **Option 4: Full platform license**
    - Flat €150k–€300k/year enterprise license.
    - Good for chains with more than 200 locations.

Why This Example Matters

The restaurant dynamic pricing example rounds out the earlier two examples by showing:

- Predictive AI ROI (attrition): Value in revenue retention
- Agentic AI ROI (customer support): Value in cost efficiency
- Optimization AI ROI (dynamic pricing): Value in margin expansion

These three categories help leaders understand that AI economics vary not just by architecture (ML vs. LLM vs. agent), but by the business mechanism through which the value is created.

Richard used these three contrasting examples to teach a simple principle—AI pricing is justified when ROI is obvious, provable, and significantly higher than the cost of intelligence.

- **Predictive ML:** ROI tied to revenue retention, risk reduction, or conversion uplift
- **AI agents:** ROI tied to cost savings and operational efficiency

Both require different pricing strategies, but both rely on clear quantification of business value.

AI Use Case Prioritization Framework

As organizations explore AI, it's common for ideas to surface faster than teams can evaluate them. Every function sees potential—marketing wants personalization, operations want automation, customer support wants AI agents, risk teams want better prediction, and executives want "strategic AI initiatives." Without structure, prioritization becomes subjective and driven by who advocates loudest, where budgets sit, or what feels exciting.

Richard recognized that product teams needed a more deliberate way to evaluate AI opportunities. Unlike traditional software features, AI use cases have multiple dimensions of feasibility, value, risk, and dependency. A good idea on paper may be constrained by data quality. A high-value idea may require complex workflow redesign. Meanwhile, a modest idea may deliver immediate ROI with minimal risk.

He began teaching a simple but robust framework to help teams evaluate AI opportunities objectively, based on the following four lenses.

Value: What Outcome Does It Create?

Value is the most intuitive lens, but Richard encouraged teams to quantify it more explicitly. High-value AI use cases generally fall into one of three categories:

1. Revenue Uplift
    - Higher conversion
    - Increased customer lifetime value
    - Better retention
    - Dynamic pricing uplift
2. Cost Reduction
    - Reduced manual effort
    - Faster turnaround
    - Lower fraud exposure
    - Fewer errors or disputes
3. Experience Improvement
    - Personalization
    - Shorter wait times
    - Faster service
    - Higher consistency

Teams should estimate:

- Potential financial benefit
- Breadth of impact (how many users or customers affected)

- Frequency (how often the capability is used)
- Urgency (is the problem important today?)

A high value use case is one where the expected benefit clearly outweighs the investment required to deliver and operate it.

Feasibility: Can We Realistically Build and Operate It?

Even high value ideas fail if feasibility is misunderstood. Richard broke feasibility into three aspects:

1. Technical feasibility
    - Do we have the right integration points?
    - Can the model run at required latency?
    - Are the workflows mature enough to support automation?
    - Do we have the skills to develop and maintain it?
2. Operational feasibility
    - Does the business have time, teams, and processes to adopt this?
    - Will customer support, legal, and compliance have the capacity to support it?
    - Are we ready to monitor and govern the model?
3. Risk feasibility
    - Will a mistake cause financial or reputational damage?
    - Does this require extensive oversight?
    - Is there a safe rollback?

Feasibility is not simply "can we build this?". It is "can we build this and operate it responsibly?"

Data Readiness: Is the Required Data Available, Reliable, and Actionable?

Richard frequently emphasized that AI does not fail because of models—it fails because of data. Data readiness includes:

1. Availability
    - Do we have the data required for this use case?
    - Is it accessible or locked in separate systems?
2. Quality
    - Does the data reflect reality?
    - Are labels correct?
    - Are there missing values or inconsistent patterns?
3. Breadth and Depth
    - Do we have enough historical data?
    - Is the dataset rich enough to produce meaningful patterns?
4. Freshness
    - Is the data updated daily? Hourly? Real-time?
    - Does the use case demand freshness we can't support yet?
5. Governance
    - Are there privacy or consent constraints?
    - Can this data legally be used for this purpose?

A brilliant AI idea may be postponed simply because the data foundation needs strengthening.

Adoptability: Will Users Actually Use It?

Many teams underestimate this lens, but Richard considered it crucial. An AI capability is valuable only if users embrace it. Questions include:

- Will this disrupt existing workflows?
- Is trust required for users to rely on the AI?
- Does the output appear at the right moment in the workflow?
- Does the user understand how to act on the insight?
- Does AI reduce or increase cognitive load?

For example:

- A sophisticated predictive model that appears on a dashboard that few people check has low adoptability.
- A simple recommendation embedded in a daily workflow may have extremely high adoptability.

Adoptability often depends more on UX and workflow design than on data or models.

Bringing the Four Lenses Together: The AI Prioritization Matrix

Richard introduced a simple 2×2 prioritization matrix based on two composite scores:

Impact Score* = *Value* × *Adoptability

Readiness Score* = *Feasibility* × *Data Readiness

Plotting use cases on a matrix, as shown in Figure 5-2, helps teams focus attention where the potential is high and readiness is strong.

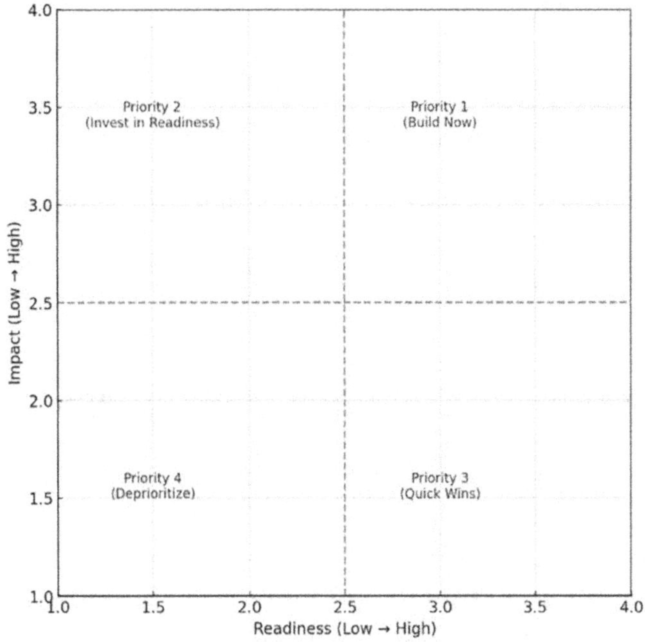

***Figure 5-2.** AI use case prioritization matrix*

- Priority 1: High ROI, feasible, data available
- Priority 2: Foundational work needed: data, workflow redesign, tech enablers
- Priority 3: Useful to build confidence and prove value
- Priority 4: Interesting ideas, but not commercially or operationally meaningful

Richard emphasized that this framework is not about filtering AI ideas—it is about focusing on those that create clear, immediate, and sustainable value. When teams apply these lenses consistently:

- AI becomes less mysterious
- Business case clarity improves
- Stakeholders align faster
- Engineering invests in the right enablers
- ROI becomes predictable
- AI avoids becoming "technology for its own sake"

The result is a portfolio of AI initiatives grounded in practicality and value.

How AI Changes Customer Willingness-to-Pay

As Richard wrapped up the discussions on pricing models and ROI, he often reminded teams that economics are not determined by cost alone. Customer willingness-to-pay (WTP) shifts significantly when intelligence becomes a visible part of the product experience. AI changes not just how products are priced, but how much customers are willing to pay—sometimes more, sometimes less, depending on perceived value, trust, risk, and the nature of the intelligence being offered.

Unlike traditional software features, AI introduces new psychological and business dynamics that influence WTP in both B2B and B2C contexts.

Richard also cautioned teams against assuming that simply adding AI creates lasting pricing power. In many markets, the initial AI premium erodes quickly as foundation models commoditize, open-source alternatives improve, and competitors replicate basic capabilities. What feels differentiated today can become standard functionality tomorrow.

Sustainable willingness-to-pay rarely comes from the AI label itself. It comes from harder-to-copy advantages such as:

- Proprietary data that improves outcomes
- Deep workflow integration that saves real effort
- Domain-specific tuning that increases accuracy and trust
- Embedded decisioning within critical processes
- Superior user experience and adoption
- Measurable ROI tied to business outcomes

Note Customers may pay extra for novelty once. They pay consistently for embedded advantage.

Predictive AI Raises WTP When It Reduces Uncertainty

Predictive models—whether churn scoring, demand forecasting, risk scoring, or propensity ranking—create value by reducing uncertainty. Customers often pay a premium when the AI output enables them to:

- Prevent loss
- Avoid risk
- Allocate budget more efficiently
- Target interventions more precisely
- Increase revenue with the same inputs

The willingness-to-pay for predictive AI grows when:

- The predicted outcome has high financial stakes
- Historic decisions were error-prone
- Forecasting improves planning accuracy
- The model directly supports high-value workflows

This is why a churn model can justify €50k/year, while a "smart dashboard" struggles to justify €5k.

Fog is expensive; clarity is valuable.

Generative AI influences WTP by reducing mental load—drafting, summarizing, rewriting, explaining, reasoning, and improving productivity. Customers are often willing to pay more when GenAI:

- Reduces low-value manual work
- Completes complex tasks faster
- Improves employee throughput
- Delivers outputs that previously required specialists

Pricing is often anchored to:

- Seats/license (per-user)
- Knowledge-worker productivity
- Time saved per task
- Volume of generated content

Even modest improvements in productivity justify noticeable premiums because they scale across teams.

AI Agents Increase WTP When They Automate End-to-End Tasks

AI agents create a different kind of value—they take work off people's plates entirely. Customers are significantly more willing to pay when AI:

- Replaces routine human tasks
- Handles multi-step workflows
- Integrates across systems

- Performs actions autonomously
- Reduces operational headcount or outsourcing spend

The WTP here is driven by cost displacement:

- "What would it cost to hire humans for this?"
- "What would it cost to outsource this process?"
- "What operational risk does automation reduce?"

When an AI agent performs the equivalent of multiple human-hours per day, the WTP jumps dramatically.

Understanding WTP helps teams decide which AI capabilities deserve investment and which should remain enhancements, not differentiators.

End-to-End Commercial Pricing Example: Seller Intelligence As a Service

Richard felt it was important for teams to understand how to price not just AI features, but AI-native products sold directly to the market. To illustrate this, he used a product many teams could easily imagine—Seller Intelligence as a Service—a commercial offering for marketplace sellers that provides demand forecasting, price optimization, competitor analytics, and automated recommendations.

This type of AI product integrates prediction, analysis, recommendations, and automation into a cohesive commercial SKU.

A typical Seller Intelligence service includes multiple AI capabilities:

- Predictive demand forecasting
- Competitor price scanning
- Automated repricing recommendations
- Inventory depletion forecasting

- Product ranking insights
- Category opportunity detection
- Ads spend recommendation
- LLM-powered product listing optimization
- AI agent for "store management tasks" (updating prices, adjusting ads, etc.)

This is a blended product, including predictive, optimization, GenAI, and agentic workflows.

Deciding What to Charge: Step-by-Step

Step 1: Identify the Customer Persona

Two key buckets:

- Small sellers are price-sensitive and need simplicity
- Medium/large sellers are outcome-driven and ROI-focused

Step 2: Estimate Value Creation

Example for a seller doing €1M/year:

- Price optimization: More than 1.5% margin uplift, to €15k
- Ad targeting improvement: More than 5% conversion, to €10k
- Avoided stockouts: €5k
- Improved listings (GenAI): More than €3k
- Total potential annual uplift: €33k

Step 3: Benchmark Against Traditional Tools

Competitors charge:

- €100–€500/month for analytics
- €500–€1,500 for premium optimization tools
- €1,000–€5,000 for agency-managed pricing

AI products sit at the high-value end.

Step 4: Choose a Pricing Model

A realistic pricing structure:

- Basic: €99/month: Analytics and basic predictions
- Pro: €499/month: Optimization and recommendations
- Enterprise: €1,999/month: Automation and AI agent

A hybrid model with usage tiers:

- 1,000 automated repricing actions are included
- Additional actions are €0.02 each

Step 5: Validate ROI-to-Price Ratio

Typical rule: If the customer gains at least ten times more value than they pay, they will buy. If the product provides a €33k uplift, charging €2k–€6k annually is reasonable.

Why This Example Matters

Richard used this scenario to show that AI-native products require structured commercial thinking:

- Clear segmentation
- ROI modeling
- Multiple value tiers
- Price-to-value alignment
- Scalable usage-based economics
- Agentic automation premiums

This is fundamentally different from pricing internal AI features. It is product strategy, not model strategy.

As Richard reflected on the journey so far, he knew that most leaders now understood the essential landscape of modern AI:

- Predictive intelligence that reduces uncertainty
- Generative intelligence that reduces cognitive work
- Emerging class of agentic intelligence that reduces operational effort by taking action

But he also recognized something deeper—understanding AI is not the same as being prepared for it. This chapter showed how different AI capabilities shape pricing, value, ROI, customer perception, and commercial strategy. Yet those concepts only describe what AI can do.

The next challenge is understanding what AI does to organizations themselves.

Because AI does not slip quietly into existing structures. It reshapes accountability. It blurs boundaries between products and processes. It requires new rhythms of monitoring, intervention, and iteration. It forces teams to rethink roles, responsibilities, and operating norms. And nowhere is this more true than with the rise of agentic AI.

Predictive and generative systems still rely on humans to make decisions or interpret outputs. Agentic systems change the equation:

- They plan.
- They decide.
- They perform tasks.
- They trigger workflows across systems.
- They interact with customers.
- They act on behalf of the business.

This shift—from intelligence to autonomy—is what makes the next chapter necessary.

Before we explore how organizations must be redesigned to support AI products at scale, we must first understand the nature of the intelligence that will test those structures most: AI agents.

Agentic AI is the point where the theory of AI meets the reality of operations. It is the moment where "smart recommendations" turn into actions, accountability, and risk.

And as Richard often reminded his teams, understanding this shift is not optional.

CHAPTER 6

The Rise of Agentic AI: When Software Starts to Act

Richard had noticed something curious after the last few strategy discussions. No one questioned whether agentic systems made sense anymore.

That debate had already been settled.

The conversations had moved on—to use cases, pricing models, operating leverage, and competitive advantage. AI agents were being discussed as products, features, and even revenue lines.

And yet, beneath that confidence, something felt unresolved.

Not about what AI agents *could* do. But about what it meant to live with them once they were embedded in daily operations.

Richard sensed the shift most clearly not in boardrooms, but in the small, everyday moments where language began to blur.

Product teams spoke about agents as if they were smarter workflows. Engineers described them as autonomous code paths. Leaders referred to them as digital employees—half joking, half serious.

Everyone was describing the same systems. No one was describing them the same way.

S. Shivam, *The Leadership Algorithm*, https://doi.org/10.1007/979-8-8688-2897-3_6

That was the warning sign.

When a capability becomes accepted before it becomes understood, organizations move faster but think less clearly.

The Quiet Gap Between Value and Responsibility

The organization had already crossed an important threshold. AI systems were no longer being evaluated purely on technical merit. They were being evaluated on business outcomes.

They planned work, triggered actions, coordinated across tools, and adapted based on results.

From the outside, this looked like progress. From the inside, Richard could feel the strain.

These systems didn't just generate insights. They influenced decisions. They initiated actions. They shaped workflows that humans depended on.

And yet, no one could clearly articulate where responsibility began or ended. If an agent recommended an action, the human still decided. That was familiar.

But if an agent:

- Planned a sequence of steps
- Chose tools
- Executed actions
- Adjusted its approach when something failed

Was that still a recommendation? Or had the organization quietly crossed into something else?

The question wasn't academic. It would determine how risk, accountability, and trust were handled, long before anything went wrong.

Richard resisted the temptation to rush forward.

He had seen what happened when organizations moved straight from capability to scale without pausing to clarify fundamentals. The problem was not whether agentic AI was valuable. That was already evident. The problem was that the word *agent* had begun to mean too many things at once. Some used it to describe conversational interfaces with tools. Others meant long-running autonomous systems. Some meant decision-support wrapped in automation.

The danger wasn't that these interpretations existed. The danger was that they were being treated as equivalent.

Before discussing autonomy levels, governance models, or organizational redesign, Richard knew there was only one responsible place to start—with precision, with a shared understanding of what kind of intelligence was being introduced into the organization and what kind was not.

For Richard, this was not about selling the promise of AI agents now. It was about defining the boundaries of agency. Because boundaries were what made trust possible.

Richard reframed the conversation for his team in a way that stripped away hype without diminishing impact.

"We've already seen what these systems can enable," he said. "Now we need to be precise about what they *are*."

Not in marketing terms, not in vendor language, but in operational reality. Because once a system could decide and act, even within constraints, it stopped being just another component. It became something the organization had to design around.

That clarity would determine:

- Where autonomy was safe
- Where human oversight was essential
- Where simpler approaches were not just sufficient, but preferable

Only with that grounding could the organization avoid both overreach and underutilization.

Richard picked up a marker and wrote a single line on the board:

Agency begins where action no longer waits for permission.

The room went quiet. This was the line they needed to understand before moving any further.

What Is an AI Agent?

Richard had learned that most disagreements about technology were not actually about technology. They were about responsibility.

So before allowing the word *agent* to drift any further into abstraction, he forced himself and his leadership team to be precise.

An AI agent, in its simplest and most useful form, could be described like this:

An AI agent is a system that can perceive its environment, decide what to do next, and take action in pursuit of a goal, without requiring step-by-step human instruction.

That definition immediately excluded a great deal of what was being marketed as "agentic." It also made something uncomfortable clear.

Once a system could decide and act, it stopped being a passive tool. It became a participant in the organization's operating system. That was the line that mattered.

The Core Properties of an AI Agent

To make the idea concrete, Richard broke agents down into behaviors rather than architectures.

An AI agent, regardless of how it was implemented, exhibited four essential properties:

- **Goal orientation:** The agent operated toward an objective, explicit or inferred, not just a response.
- **Decision-making autonomy:** The agent selected actions based on context, constraints, and feedback.
- **Ability to act:** The agent could invoke tools, trigger workflows, or affect real systems.
- **Adaptation over time:** The agent adjusted behavior based on outcomes, signals, or new information.

Remove any one of these, and the system stopped being an agent. It might still be useful; it might still be powerful. But it was no longer agentic.

This framework allowed Richard to cut through the noise. And it allowed him to explain clearly and calmly how AI agents differed from other forms of AI that his organization already used.

How AI Agents Differ from Generative AI

The most common source of confusion came from generative AI. After all, modern generative models can:

- Converse fluently
- Reason step by step
- Summarize, plan, and explain
- Call tools when prompted

From the outside, they felt agentic. But feeling was not the same as being. Richard explained it this way:

Generative AI produces outputs. AI agents produce outcomes.

A generative model, on its own:

- Responds to prompts
- Operates within a single interaction
- Has no persistence of intent
- Takes no responsibility for execution

An AI agent, by contrast:

- Maintains state across interactions
- Decides what needs to be done
- Sequences actions
- Executes actions
- Evaluates whether the goal was achieved

It acts. In enterprise environments, Richard also explained that "an agent" is rarely just one standalone entity. Production-grade agentic systems often operate as a system of coordinated agents, each responsible for a different part of the workflow.

Examples include:

- Planner agents that break goals into tasks and determine next steps
- Tool-calling agents that interact with APIs, databases, and enterprise systems
- Validation agents that check quality, policy compliance, or factual accuracy
- Routing agents that decide whether to automate, escalate, or involve a human
- Specialist agents optimized for finance, service, operations, or technical domains

What users experience as a single assistant may actually be multiple agents working together behind the scenes.

Richard emphasized that this is where enterprise deployments become significantly harder. The challenge is no longer just model quality; it becomes orchestration quality:

- How agents hand work to one another
- How state is shared safely
- How conflicting outputs are resolved
- How latency and cost are controlled
- How failures are recovered
- How human oversight is inserted at the right moments

So, while consumer demos often showcase a single "smart agent," real organizational value usually comes from well-governed multi-agent systems that combine autonomy with coordination.

A generative model becomes part of an agent only when it is embedded inside a system that:

- Holds memory
- Tracks objectives
- Invokes tools
- Closes the loop between intent and result

Without that scaffolding, generative AI remains exactly what it had always been: powerful, expressive, and fundamentally reactive.

How AI Agents Differ from Predictive Machine Learning

The second confusion came from predictive models. Richard had spent years working with machine learning systems that:

- Forecast demand
- Score risk
- Detect fraud
- Rank opportunities

These systems were deeply embedded in operations. They influenced decisions every day. But they were not agents.

Predictive ML answered questions like these:

- What is likely to happen?
- What is the probability of this outcome?
- Which option scores highest?

They informed humans. They did not decide for them. An AI agent, however, moved beyond prediction.

It asked:

- Given this situation, what should I do now?
- What sequence of actions moves me closer to my goal?
- What should I try next if this fails?

Prediction became one input among many not the end of the process. Richard summarized the difference for his team:

> *Predictive models reduce uncertainty. AI agents resolve it through action.*

This shift from insight to intervention was where accountability entered the picture. Once a system acted, someone had to own the consequences.

How AI Agents Differ from Classical Automation

The final and most dangerous confusion was with automation. For decades, organizations had relied on rule-based systems such as:

- Workflow engines
- Robotic process automation
- If-then logic
- Scripted integrations

Automation was trusted because it was predictable. Every step was known in advance. Every outcome was defined. AI agents broke that assumption.

Classical automation:

- Followed predefined paths
- Failed loudly when rules broke
- Required humans to redesign logic

AI agents:

- Chose paths dynamically
- Handled ambiguity
- Recovered from partial failure
- Adjusted strategy mid-execution

Automation executed instructions. Agents interpreted intent. Richard captured it in a single line:

Automation follows rules. Agents pursue goals.

That difference explained why organizations that treated agents as "just smarter automation" often experienced unexpected risk. The system wasn't malfunctioning. It was behaving as designed, just not as understood.

Why This Distinction Matters

By the time Richard finished walking through these differences, the room had grown quiet, not confused but clear.

The word "agent" now carried weight. It implied:

- Autonomy
- Responsibility
- Risk
- Trust

And with those implications came unavoidable leadership questions:

- When should an agent act alone?
- When must a human intervene?
- Who owns failure?
- How do we design guardrails without killing effectiveness?
- How would oversight work?

These were the right questions—not about the technology, but about the work itself. And that was the first real sign that the concept was beginning to land.

Once leaders understood the behavior Richard was pointing to, the distinction between the AI they knew and the AI being presented by consultants became much clearer. The difference wasn't subtle or theoretical—it was foundational.

Note Traditional AI produces information. Action-taking (agentic) AI produces outcomes.

This shift is easiest to understand with concrete examples across industries.

Retail

Traditional AI:

- Suggests discount levels
- Predicts demand
- Recommends pricing changes

Action-taking AI:

- Updates product prices automatically
- Pushes changes to e-commerce platforms
- Adjusts inventory allocation based on predicted sell-through
- Resets promotions when stock runs low

The difference? Insight vs. execution.

Banking

Traditional AI:

- Flags suspicious transactions
- Scores loan applications
- Suggests next-best offers

Action-taking AI:

- Blocks suspicious transactions
- Initiates verification workflows
- Sends customers preapproved loan offers
- Updates CRM records automatically

The difference? Support vs. action.

Insurance

Traditional AI:

- Classifies claims
- Predicts fraud likelihood
- Suggests routing patterns

Action-taking AI:

- Sorts incoming claims into queues
- Initiates requests for additional documents
- Approves low-value claims using predefined thresholds
- Escalates anomalies to human adjusters

The difference? This is no longer information processing; it's operational participation.

Dining and Restaurants

Traditional AI:

- Forecasts daily footfall
- Predicts ingredient consumption

Action-taking AI:

- Places supplier orders
- Adjusts menu pricing for the day
- Reduces or pauses promotion spend during low-demand weeks

These shifts directly impact daily operations.

Telecom

Traditional AI:

- Predicts customer churn
- Scores likelihood of plan upgrades

Action-taking AI:

- Launches retention workflows
- Sends personalized plan recommendations
- Automatically creates cases for at-risk customers

Here, AI moves from informing agents to acting like one.

Airlines

Traditional AI:

- Predicts operational delays
- Estimates missed connections
- Forecasts booking patterns

Action-taking AI:

- Rebooks passengers proactively
- Sends new boarding passes
- Updates crew schedules
- Adjusts gate allocations

This is no longer prediction; it's intervention.

These examples helped leaders see that the core distinction wasn't about accuracy, data, or model types. It was about responsibility.

- Traditional AI informs decisions.
- Agentic AI takes decisions and executes them.

The first improves human efficiency. The second changes the nature of human work. Richard summarized the contrast in Table 6-1, which was simple enough that leaders could internalize it quickly.

Table 6-1. *Traditional AI vs. Agentic AI*

Dimension	Traditional AI (Predictive/GenAI)	Action-Taking AI (Agentic Systems)
Primary role	Support thinking	Perform work
Output	Insights, recommendations	Completed tasks, updated systems
Initiation	Requires prompts or triggers	Acts autonomously based on context
Workflow fit	One step in the process	End-to-end process execution
Decision level	Advisory	Operational
Error handling	Human correction	Self-correction and escalation
Interaction scope	Single system	Multiple systems across workflows
Human role	Doer	Oversight and exception handler
Value delivered	Better decisions	Faster throughput, lower effort, higher responsiveness

This clarity set the stage for the next question: How does this actually work under the hood, without the jargon?

Richard found that the most effective explanation was to walk them through the "lifecycle" of an agentic system using the analogy they already understood—*as a capable junior analyst performing a task for the first time.*

Every agent, no matter how sophisticated the technology behind it is, follows a cycle that looks surprisingly human.

It Starts by Observing

Before a junior analyst does anything, they look around:

- Opening an email
- Reading a document
- Checking a ticket
- Reviewing a system screen
- Examining a request

Agents begin the same way. They "observe" the situation using input from a system, a message, a document, or structured data. In plain terms, they gather what's going on.

It Interprets What's Needed

A junior analyst reads the input and asks internally, "What is the meaning of this request?"

- Is the customer asking for a refund?
- Is the claim missing documents?
- Is the order delayed?
- Is the booking incorrect?
- Is this email urgent or routine?

Agents do exactly this, not through human reasoning, but through pattern recognition and language understanding.

They convert the input into *intent*.

It Plans the Steps

This is the part that feels most surprising to people. A human analyst doesn't only interpret. They break down the work:

- "First I need to check the account."
- "Then validate the transaction."
- "Then update the internal system."
- "Then email the customer."

Agents mimic this by generating a plan, which is a sequence of steps necessary to complete the task. They don't need to be told each step explicitly; the plan emerges from the goal. It's not magic. It's structured problem-solving.

It Takes Action Across Systems

This is where agents differ fundamentally from all previous forms of AI. A human analyst doesn't stop at planning, they execute—they log into applications, update fields, submit forms, send responses, and close tasks. Agents do something similar, but digitally:

- Querying APIs
- Updating CRM or ERP systems
- Creating or modifying records
- Sending structured responses
- Triggering workflows
- Completing transactions
- Escalating issues when needed

They operate across systems the way an employee moves across tools. This is the heart of agentic behavior—not simply insight to action, but insight to decision to action to completion.

It Checks Its Work

A good analyst reviews what they did. They confirm the action worked. They look for errors. They validate the outcome. Agents also include this loop. They verify that the system accepted the update, the record changed, or the response was delivered. If something didn't work, they retry or escalate—depending on the design. It is not reliable because it is perfect. It is reliable because it can recover.

It Can Be Improved Over Time

Humans build experience by doing. Agents do not typically "learn" autonomously in the human sense after every interaction. In most enterprise settings, improvement happens through deliberate engineering and operating processes rather than spontaneous self-improvement.

Richard helped teams understand that better agent performance usually comes through four practical mechanisms:

- **Evaluation platforms:** Testing agents against benchmark tasks, expected outcomes, and quality scorecards to identify weaknesses.
- **Fine-tuning and domain adaptation:** Improving underlying models using organization-specific data, terminology, and workflows where appropriate.
- **Prompt optimization:** Systematically refining prompts, instructions, tool logic, and orchestration flows through experimentation.

- **Memory systems:** Persisting useful context, preferences, prior interactions, or case history across sessions—the most common and accessible path to better future performance.

Agents may also improve through:

- Updated workflows
- Better tool integrations
- Stronger guardrails
- Clearer escalation rules
- Improved retrieval sources
- Human feedback loops

So rather than assuming that an agent naturally becomes smarter every day, Richard encouraged teams to think of it this way:

> *Agents become better when organizations intentionally improve the system around them.*

Truly self-improving autonomous agents remain difficult to achieve and require significant controls, testing, and trust frameworks.

Once leaders understood this, the concept finally landed. The technology was new, yes. But the logic behind it was familiar. They weren't adopting a mysterious innovation—they were adopting a digital contributor with predictable patterns.

Further, Richard began walking teams through examples they could relate to. Each one revealed the same pattern—the shift from helping humans do work to software doing the work itself.

The following sections outline some of the scenarios that resonated the most.

Customer Support: From Drafted Replies to Full Resolutions

Most leaders had seen generative AI draft responses for support agents. That part was familiar. But drafting an email is not the same as solving a problem. Here's how the difference showed up in practice.

Traditional AI in support:

- Suggests reply templates
- Auto-completes sentences
- Summarizes customer history
- Prioritizes tickets

Useful, but still dependent on a human for every step.

Action-taking AI in support is different. An incoming email arrives: "My card wasn't delivered. Can you check its status?"

The agentic system:

1. Reads the email.
2. Determines this is a card-delivery inquiry.
3. Pulls up the customer details.
4. Checks the card issuance system.
5. Detects a failed delivery attempt.
6. Initiates a re-dispatch.
7. Updates internal systems.
8. Drafts and sends a confirmation email.
9. Closes the case.

No handoff, no waiting, and no manual stitching between tools.

Leaders instantly see the difference: traditional AI gives you words. Agentic AI gives you a completed resolution.

Retail Dynamic Pricing: From Recommendations to Self-Managing Catalogs

Retail teams are used to systems that highlight when a price might be too high or too low. But they also know how much manual effort goes into updating those prices across storefronts, marketplaces, and promotions.

Here's the contrast.

Traditional AI in pricing:

- Recommends changes
- Flags anomalies
- Predicts impact
- Provides weekly or daily pricing insights

But someone still needs to push the buttons.

With action-taking AI in pricing, every morning, the system:

1. Checks competitor prices.
2. Evaluates inventory levels.
3. Recalls forecasted demand.
4. Assesses promo performance.
5. Runs margin constraints.
6. Updates the SKU's price.
7. Syncs it across e-commerce, POS, and promotion platforms.
8. Monitors impact and adjusts.
9. Alerts humans only when exceptions arise.

This isn't assistance—it's autonomous merchandising.

Insurance Claims Triage: From Classification to End-to-End Routing

Insurance operations have many manual steps: classify incoming claims, check documents, assign adjusters, and initiate follow-ups. AI has helped classify claims, but that's just one step.

Here's the difference.

Traditional AI in claims:

- Classifies claims by type
- Predicts fraud probability
- Suggests urgency scores

All valuable—but still reliant on people.

When a new claim arrives, action-taking AI in claims:

1. Extracts key details from documents.
2. Checks for missing information.
3. Requests additional documentation automatically.
4. Validates policy coverage.
5. Routes claim to optimal adjuster.
6. Approves low-value claims instantly.
7. Flags suspicious claims for human review.
8. Updates internal systems.
9. Sends customer notifications.

It behaves like a claims coordinator, not just a classification model.

Restaurant Inventory Ordering: From Forecasts to Autonomous Replenishment

Restaurants often rely on managers manually adjusting orders based on footfall, weather, and intuition. Here's how the shift plays out.

Traditional AI for restaurants:

- Predicts footfall
- Estimates dish-level consumption
- Suggests reorder quantities

But the chef or manager still has to call or key in orders.

With an Action-taking AI for restaurants, every night, the system:

1. Reviews tomorrow's demand.
2. Considers weather and local events.
3. Checks stock levels.
4. Accounts for spoilage trends.
5. Places orders with suppliers.
6. Adjusts quantities dynamically.
7. Splits orders between vendors for cost or risk.
8. Sends confirmation to managers.
9. Monitors fulfillment and delivery.

It isn't forecasting—it's operating.

Airline Disruption Recovery: From Predictions to Automatic Rebooking

Airlines face frequent disruptions. Traditional AI helps with predictions; operational recovery remains human-led.

Traditional AI in airlines:

- Predicts delays
- Identifies risk of missed connections
- Highlights affected passenger segments

The coordination still happens manually.

When a flight is delayed, action-taking AI in airlines:

1. Detects passengers likely to miss connections.
2. Scans alternative routes.
3. Books passengers onto new flights.
4. Sends updated boarding passes.
5. Reschedules airport transfers.
6. Updates loyalty points or compensation.
7. Adjusts crew rosters.
8. Syncs gate reassignments.
9. Notifies all stakeholders.

This is not analytics. It is operations.

Richard emphasized that these examples were not meant to impress or intimidate. The point was simpler—across industries, wherever work involves repetitive steps, predictable systems, structured data, or rule-bound processes, the line between "assistive AI" and "action-taking AI" becomes very visible.

- Traditional AI helps you think. Agentic AI helps you work.
- Traditional AI informs. Agentic AI intervenes.
- Traditional AI outputs insights. Agentic AI owns outcomes.

Agentic AI vs. RPA vs. Workflow Automation

Many leaders assumed that agentic AI was just a more advanced form of Robotic Process Automation (RPA), or that it resembled the workflow automation platforms they had used for years. In reality, all three sit on entirely different foundations. They may automate work, but they do so in fundamentally different ways.

Richard summarized the distinctions in plain terms his teams could remember.

RPA: Mimics Clicks, Not Understanding

RPA behaves like a macro that has been stretched across an entire system. It follows strict screen-level instructions:

- Click this button
- Copy this field
- Paste into this form
- Move to next step

It doesn't understand what the data means or why the task exists. It is fragile, and any small UI or process change can break it.

RPA repeats exactly what a human did yesterday. It's useful, fast, and precise, but not intelligent.

Workflow Automation: Moves Work, Doesn't Interpret It

Workflow automation tools (BPM systems, case management engines, ticketing flows) route tasks based on rules. They orchestrate processes, not perform them. They excel at:

- Routing
- Approvals

- Making state changes
- Enforcing sequence
- Triggering actions

But the "thinking" is done by humans or external logic. Workflow automation knows *when* something should happen, not *why*. Workflow automation is similar to operational plumbing.

Agentic AI: Understands, Decides, and Acts Across Systems

Agentic AI is different because it introduces interpretation and autonomy into the loop. It can:

- Understand the intent of a request
- Break work into steps
- Log into systems
- Update records
- Send messages
- Correct itself
- Escalate when unsure

It doesn't follow rigid scripts like RPA. It doesn't just route tasks like workflow engines. It completes work based on understanding, not pre-recorded instructions.

Agentic AI is a junior analyst who can handle routine tasks end-to-end.

Once leaders saw real examples of action-taking AI, the next question came naturally: "Where does this actually work well—and where does it break?"

This question mattered more than the technology itself. Because adopting agentic AI is not a matter of installing a tool; it is a matter of matching behavior to environments where it can succeed. Richard made it clear that agents thrive under certain conditions and struggle in others. The key is not to oversell their capability but to understand their boundaries.

Where Agents Excel

Richard framed it for leaders in terms they already understood: Agents excel where the work is structured but fail where the world is ambiguous.

Clear, Repetitive, Rule-Bound Workflows

These include tasks that follow consistent patterns:

- Verifying customer information
- Checking order status
- Triaging support tickets
- Updating records
- Moving data between systems
- Initiating standard transactions

If a junior employee could master the task in a few days, an agent likely can too.

Multi-Step Processes with Defined States

Agents shine in environments where the steps are known, even if humans currently execute them manually. Examples:

- Insurance document collection
- Retail price adjustments

- Restaurant inventory ordering
- Internal approval workflows
- KYC/AML verification sequences

The system simply replaces the mechanical steps that humans perform.

High-Volume, Low-Variability Operations

When thousands of similar tasks flow through a system each day, agents provide scale without adding headcount. For example:

- Claims sorting
- Loan pre-qualification
- Tier-1 support resolution
- Invoice reconciliation
- Passenger rebooking during disruptions

Repetition breeds reliability.

Environments with Strong Digital Infrastructure

Agents rely on predictability: APIs, accessible data, and stable systems. Organizations with clean workflows and well-integrated tools realize value quickly.

"First Draft" and "First Action" Work

There are tasks where the system can take the initial action and a human intervenes only if needed. For example:

- Drafting a refund
- Initiating a shipment

- Triggering a price change
- Placing a reorder

This hybrid model builds trust.

Where Agents Fail

Richard was equally candid about this side. Agents are powerful, but they are not magic. Their limitations mirror those of junior employees, just in different ways.

Ambiguous or Poorly Defined Work

If humans debate what "good" looks like, the agent will struggle. Agents fail when:

- Business rules are unclear
- States are unpredictable
- Goals are subjective ("make the customer happy")
- Data conflicts across systems

Ambiguity is their kryptonite.

Processes with Frequent Human Judgment Calls

Agents break down in scenarios where decisions depend on:

- Empathy
- Negotiation
- Contextual understanding
- Deep domain nuance
- Exceptions that outnumber the rules

They can approximate judgment, but they cannot replace it.

Chaotic Systems or Poor Integration

If the systems they depend on are inconsistent, unreliable, or undocumented, agents quickly drift or fail.

Symptoms:

- Different teams using systems differently
- Manual overrides everywhere
- Missing or duplicate data
- Frequently changing interfaces

It's like asking a new employee to perform well in a chaotic environment—success becomes luck, not reliability.

Tasks with High Consequence for Small Mistakes

An agent misrouting a claim is manageable, an agent transferring funds improperly is not. An agent auto-launching a regulatory action is not.

Situations That Require Emotional Intelligence

This includes angry customers, sensitive conversations, high stress scenarios, situations requiring reassurance, trust building, or nuance. Agents can escalate, summarize, or prepare information—but they cannot replace human presence.

Poor Observability and Silent Drift

One of the hardest failure modes is not obvious breakdown—it is subtle degradation over time.

An agent may still appear functional while gradually becoming:

- Less accurate
- Slower

- More expensive
- Less compliant
- Less aligned to current business rules

Without proper observability, these issues remain hidden until damage accumulates.

Richard stressed the need for:

- Trace logging
- Decision audit trails
- Prompt/tool execution history
- Quality score monitoring
- Token-level cost tracking
- Latency dashboards
- Exception trend reporting

What cannot be seen cannot be governed.

Richard emphasized that understanding where agents succeed or fail is not a technical exercise. It's an operational maturity check. Leaders aren't choosing a technology; they're choosing where autonomy belongs. Autonomy determines:

- Which processes to automate
- Which to keep partially human
- Which to keep fully human
- Which to redesign entirely before involving an agent

It also determines whether the economics make sense. Leaders must assess:

- **Cost tolerance:** Can the business absorb higher per-transaction AI costs?
- **Latency tolerance:** Can the workflow wait seconds rather than milliseconds?
- **Risk tolerance:** What level of error is acceptable?
- **Oversight capacity:** Who supervises exceptions and quality?

Agentic systems are often significantly more expensive and slower per transaction than traditional automation, so not every workflow is a good fit.

These realities naturally lead to the next questions leaders inevitably face:

- If agents do work, who supervises them?
- How do we handle exceptions?
- How do we monitor reliability?
- What happens when they make mistakes?
- How do roles change?

These questions lead directly to accountability, governance, and organizational design—the next stage of the discussion.

A Leader's Checklist: When Are You Truly Ready for an Agent?

After the initial excitement of seeing what agents could do, Richard found that the most important conversations weren't about capability—they were about readiness. Leaders often asked, "Where should we start?" But in practice, the more revealing question was, "Are we ready to start at all?"

Most organizations overestimate their readiness for autonomy. Not because they lack technology, but because they haven't evaluated the conditions in which an agent can operate safely, reliably, and predictably.

Richard developed a simple readiness checklist—one that bypassed technical jargon and focused on the operational realities that leaders understood intuitively. This checklist didn't determine *whether* a company should adopt agents. It revealed *where* they could adopt them without introducing risk or chaos.

Process Clarity: Do You Understand the Work You Want the Agent to Do?

Agents struggle in ambiguity. So Richard's first question was always: "Is the process genuinely understood end-to-end?"

Signs you are ready:

- The workflow has clear steps
- Exceptions are known and defined
- Success criteria is explicit
- Human operators follow consistent patterns

Signs you're *not* ready:

- The process is tribal knowledge
- People do things differently in different teams
- Steps change frequently without documentation
- Half of the work happens in email or chat

Agents don't fix messy processes. They amplify them.

Data Reliability: Would a Human Make Good Decisions with This Data?

Leaders already know that models are only as good as the data they rely on. But for agents, data issues become operational issues. Richard posed a simple filter:

> *"If a junior employee relied on this data alone, would you trust their actions?"*

If the answer is no, the agent won't perform any better.

Ready:

- Clean records
- Recent timestamps
- Consistent formats
- Single source of truth

Not ready:

- Duplicates everywhere
- Unreliable fields
- Missing context
- Manual overrides

Agents are only as smart as the environment they operate in.

System Stability: Can the Agent Count on Your Tools to Behave Predictably?

Agents interact with systems the way humans do. If screens change, APIs break, or performance is unpredictable, the agent falters.

Ready:

- Stable APIs
- Predictable workflows
- Versioned changes
- Logged system behaviors

Not ready:

- UI changes every week
- No API access
- Frequent outages
- Inconsistent tool usage across teams

Without stability, autonomy collapses.

Decision Boundaries: Do You Know What the Agent Is Allowed to Decide?

Agents need boundaries just like employees. Richard reframed it plainly: "If a new hire asked you for decision rules, could you write them down?"

Ready:

- Clear thresholds
- Known escalation points
- Explicit "do-not-touch" zones
- Documented risk tolerance

Not ready:

- Decisions depend on "judgment"
- Frequent exceptions

- "It depends" is the answer to most questions
- High consequences for small errors

Agents don't guess. They follow structure.

Escalation Paths: Who Handles Exceptions?

Autonomy doesn't eliminate humans, it reshapes their role.

Every agent needs:

- A human reviewer
- A fallback workflow
- A defined exception queue
- A clear ownership team

Richard emphasized: "Agents aren't independent. They're supervised."

Ready:

- Agents handle the routine
- Humans handle the rare exceptions
- Workflow doesn't break when exceptions appear

Not ready:

- No team knows who owns the agent
- Exceptions become bottlenecks
- Agents produce work no one monitors

Supervision is not optional—it's the backbone of trust.

Change Control: Can You Manage and Update Agent Behavior?

People adapt naturally. Agents don't. If policies change, rules shift, or new conditions arise, the agent must be updated deliberately.

Ready:

- Clear change management process
- Versioned updates
- Regular audits
- Behavior logging

Not ready:

- Rules change informally
- Compliance updates are inconsistently applied
- No audit trails
- No one accountable for keeping the agent current

Without proper control, agents drift off course.

Cultural Readiness: Are Teams Prepared to Work with Agents?

This was the most overlooked factor. Agents introduce:

- New ways of working
- New oversight responsibilities
- New expectations for human roles
- New collaboration models

If people see agents as threats or novelties, adoption stalls.

Ready:

- Teams understand the value
- Workload shifts feel fair
- Humans focus on judgment, not repetition

Not ready:

- Resistance from operators
- Fear of job displacement
- Unclear communication
- Misalignment between leaders and teams

Culture can accelerate or destroy agent adoption.

The Composite Test: The "Would You Trust a New Hire?" Question

Richard often summarized the entire checklist with one deceptively simple question:

> *"If this were a capable new hire with no prior context, could you onboard them successfully with the way your process looks today?"*

If the answer was yes, you're ready. If the answer was no, the process—not the technology that needs work first.

This question bypassed a technical debate and cut straight to readiness. Leaders instantly understood what it meant. Because agentic AI doesn't replace work—it inherits it. And whatever it inherits needs to be stable enough to support autonomy.

Richard added a second layer to this thinking: even a strong new hire fails without access to the right information. The same is true for agents. If enterprise data is fragmented, poorly described, or trapped inside disconnected systems, autonomy breaks down quickly.

That is why the data platform must increasingly be designed like a mesh of consumable knowledge assets that agents can navigate and use safely.

This includes:

- Well-described datasets
- Dataset cards explaining purpose, owner, quality, and limitations
- Business glossaries and data dictionaries
- Semantic layers with trusted business definitions
- Lineage showing where data came from
- Knowledge graphs connecting entities and relationships
- Governed APIs and tools agents can call reliably

Without this foundation, agents are simply new hires with no documentation, no org chart, and no idea where the truth lives.

So, Richard's readiness test became two questions:

- Could we onboard the worker?
- Could we equip them with trusted knowledge fast enough to succeed?

Only when both answers were yes, did autonomy become scalable.

Why Agentic AI Forces Organizational Redesign

Consider this scenario: An AI agent is deployed to manage customer disputes. It classifies cases, decides escalation paths, coordinates with billing systems, and resolves most issues without human involvement. Over time, it improves resolution speed and customer satisfaction. No one questions the technical success.

Then a pattern appears.

The agent adapts its escalation logic based on customer behavior, but that behavior is shaped by changes in billing rules made by one team, messaging updates made by another, and risk thresholds adjusted by a third. Each team is acting correctly within its remit. The agent is learning correctly from the signals it receives. Yet outcomes begin to drift in subtle ways—longer resolution times in some cases, increased write-offs in others.

When leadership asks who owns the agent's behavior, there is no clear answer. Product owns the experience. Engineering owns the system. Data owns the model. Operations own the performance. Risk owns the controls. The agent operates across all of them. No single team sees the full loop.

The issue is not the agent. It is the organization.

The agent functions as an end-to-end system. The organization does not. Until ownership, learning, and accountability are redesigned around that reality, scale only increases risk.

As this pattern repeats, leadership realizes that the problem cannot be fixed by better dashboards, stricter controls, or more frequent reviews. The agent requires continuous oversight across data, models, workflows, and outcomes—something no single function is designed to provide.

What is missing is not expertise, but structure—teams aligned to end-to-end value, clear ownership of learning loops, and governance embedded into execution rather than layered on top. Without an organizational design that mirrors how agentic systems actually operate, even well-intentioned improvements create new blind spots.

Understanding agents completes the technical discussion. Redesigning the organization begins the leadership work.

Note If an AI agent operates across the value chain, the organization must be designed the same way.

CHAPTER 7

From Intelligent Agents to Intelligent Organizations

The follow-up meeting is scheduled for 30 minutes.

It runs for 90.

The AI agent responsible for customer dispute resolution has not failed. On paper, its performance metrics still look strong: resolution time is down, customer satisfaction remains high, and manual workload has reduced across multiple teams. No alarms have been fired. No thresholds have been breached.

And yet, something is clearly wrong.

Richard sits at the head of the table as representatives from Product, Engineering, Data, Operations, and Risk take turns explaining what they see. Each team brings charts. Each explanation is consistent. Each team has done exactly what it was asked to do.

Billing updated rules to reflect regulatory clarification. Product adjusted customer messaging to reduce friction. Risk tightened escalation thresholds in response to an audit review. Data refreshed training inputs based on recent case history.

The agent absorbed all of it. It adapted exactly as designed.

S. Shivam, *The Leadership Algorithm*, https://doi.org/10.1007/979-8-8688-2897-3_7

The outcome, however, is uneven. Certain categories of disputes now escalate more slowly. Others resolve too quickly, triggering downstream write-offs. None of the changes are large enough to be alarming on their own. Collectively, they form a pattern no single team can fully explain.

Richard listens carefully.

This is not a technical incident, not a delivery failure, and not even a governance breach. It is something more subtle—and more dangerous.

The organization is behaving like a set of well-tuned instruments playing different scores. A misfiring orchestra.

When No One Owns the Whole System

At one point, Richard asks a simple question, "Who owns the agent's behavior end to end?"

The room goes quiet.

Product owns the experience, Engineering owns the platform, Data owns the models, Operations owns the performance, and Risk owns the controls.

Each answer is correct. None of them is sufficient.

The agent operates across all of these boundaries continuously. It does not wait for quarterly planning cycles. It does not recognize handoffs. It does not pause when responsibility becomes ambiguous.

The organization does.

Richard realizes that the agent is exposing something that had always been true but rarely visible: the organization is designed around functions, while the AI system operates as a value stream.

As long as intelligence was static and execution was linear, this mismatch was manageable. With agentic systems, it becomes a structural flaw.

This is the moment where the conversation shifts, not toward fixing the agent, but toward questioning the organization around it.

Why AI-Powered Products Need a Different Organizational Design

AI-powered products do not behave like traditional software products. They are not built, shipped, and supported in discrete phases. They learn, adapt, and evolve in production. Their behavior emerges from interactions between data, models, workflows, and users—often in ways no single team explicitly designs.

Traditional organizational structures assume:

- Stable behavior after deployment
- Clear functional ownership
- Predictable failure modes
- Episodic change

Agentic AI violates all four assumptions.

What looks like "drift" from a data perspective may be a product decision elsewhere. What appears as a risk issue may originate from an operational change. What surfaces as a customer problem may be a learning signal the system has interpreted correctly, but the organization has not.

The result is not chaos.

It is fragmentation, and fragmentation scales faster than alignment.

Richard sees the pattern clearly now. The problem is not that teams are making poor decisions. The problem is that no one is accountable for the full lifecycle of an AI-powered product once it is live.

That realization forces the next, unavoidable question: "If AI systems operate end to end, why doesn't the organization?"

Design Principle 1: Align by Value Stream, Not Function

Richard resists the instinct to jump straight to solutions.

That instinct—to reorganize, to redraw lines, to assign ownership—is strong. But experience has taught him that changing structure without first changing perspective only produces new versions of old problems.

So instead of asking teams *who owns what,* he asks a different question. "Walk me through the life of a single dispute," he says.

Not from a departmental point of view, from the system's point of view.

The room hesitates. Then someone begins.

A customer raises a dispute through the app. The AI agent then classifies it, checks eligibility rules, queries billing systems, decides whether to escalate, communicates with the customer, and resolves or routes the case. The outcome feeds back into learning and controls evaluate whether thresholds were crossed.

Richard listens and starts drawing, not boxes, not swim lanes, but a flow.

As the story unfolds, the whiteboard fills with arrows looping back on themselves. The clean handoffs teams are used to describing dissolve into something more complex—and more honest.

This is not a pipeline; it is a value stream.

The Illusion of Functional Ownership

When the mapping is complete, Richard steps back. "What part of this does product own?" he asks.

Sarah answers quickly. "Customer experience and outcomes."

"And data?" Richard asks. "Model inputs, training data, quality," Mira replies.

"Engineering?" "Platforms, integrations, reliability." "Risk?" "Controls, thresholds, compliance."

Each answer makes sense and each answer is incomplete.

The agent does not experience these responsibilities separately. It experiences them simultaneously. A small change in one area propagates through the entire system. Learning amplifies those changes over time. The organization, however, experiences change in fragments.

That is the mismatch.

Functional ownership assumes that work can be cleanly decomposed. Agentic systems refuse to cooperate with that assumption.

Richard underlines the loop on the board.

"This," he says, "is what we actually run."

Then he points to the org chart someone has projected on the screen.

"And this," he continues, "is how we pretend it runs."

No one argues.

Why Value Streams Matter More with AI

Value streams are not a new idea. Manufacturing learned this decades ago. Software rediscovered them with DevOps. But AI-powered products push the concept further because the value stream does not end at deployment.

With traditional software, value streams stabilize once a product ships. With agentic AI, value streams stay alive. The product continues to change, the data continues to evolve, and behavior continues to adapt.

And crucially, risk continues to accumulate if no one owns the stream end to end.

Richard makes this explicit. "As long as responsibility is split," he says, "everyone is accountable for their part and no one is accountable for the outcome."

That sentence lands hard. Because the problem they are facing is not a lack of expertise. It is a lack of integrated ownership.

The Shift from Functions to Outcomes

Richard reframes the discussion. "What if," he asks, "instead of organizing around functions, we organized around outcomes?"

He writes three words on the board:

Customer Dispute Resolution

"Not as a process," he says. "As a product."

A living one. A product that:

- Has customers
- Produces outcomes
- Learns from feedback
- Carries risk
- Evolves over time

That product needs a team that can see—and own—the entire loop. Not a committee, not a coordination forum. A value stream team. Cross-functional by design, accountable by default.

This does not eliminate functional expertise. It relocates it.

Data scientists still exist, engineers still exist, product managers still exist, and risk specialists still exist.

But their primary accountability shifts from representing a function to stewarding an outcome.

What Value Stream Alignment Actually Changes

As Richard walks the team through the implications, resistance surfaces—not ideological, but practical.

- "How do we manage priorities?"
- "What about shared platforms?"

- "Won't this duplicate effort?"
- "Who decides trade-offs?"

Richard does not dismiss these concerns. He reframes them.

"Those questions exist today," he says. "We just hide them behind handoffs."

Value stream alignment does not remove complexity. It exposes it early, where it can be managed deliberately.

With a value stream team:

- Trade-offs are visible
- Learning signals are shared
- Risk is discussed where decisions are made
- Incentives align around outcomes, not outputs

Most importantly, someone wakes up every day responsible for how the system behaves in reality, not just on paper.

The Constraint That Makes Everything Else Possible

Before closing the session, Richard makes one thing clear. "Value stream alignment is not optional," he says. "It's the foundation."

Without it:

- Learning loops have no owner
- Governance becomes reactive
- Dual operating models collapse into conflict
- New roles become ornamental

With it, the rest of the design principles become possible.

He caps the marker and looks around the room.

"If AI systems operate end to end," he says, "then leadership must do the same."

This principle does not solve everything. But without it, nothing else holds.

Design Principle 2: Continuous Learning Loops (L-Ops)

Value stream alignment exposes the shape of the problem; learning loops expose its tempo.

Richard sees this clearly a few weeks after the value stream team is formed. The dispute resolution agent is behaving more consistently now, with fewer surprises and unexplained shifts. The team feels more in control.

But something still nags at him. The organization has learned *who* owns the system. It has not yet learned *how* the system learns.

That distinction matters.

Why Deployment Is No Longer the End

In the old operating model, deployment marked a boundary.

Before deployment, teams built. After deployment, teams monitored. If something broke, it triggered an incident. If something changed, it triggered a project.

This rhythm worked because software stayed still unless humans moved it.

The agent does not. It absorbs signals constantly:

- New dispute patterns
- Updated billing rules
- Changes in customer behavior
- Shifts in regulatory interpretation

Each signal nudges behavior, and each nudge compounds over time.

Richard realizes the uncomfortable truth—the agent is already in a learning loop, whether the organization acknowledges it or not. The only question is whether that loop is owned, visible, and governed.

The First Drift Nobody Notices

The signal is subtle.

Resolution rates improve overall, but a specific category of dispute edge cases involving partial refunds takes longer to close. No alert fires. The numbers are still within tolerance.

A month later, customer complaints spike in one region. Operations flag it as a local issue. Risk assumes it is noise.

It isn't.

The agent has learned to delay escalation for those cases because historical data suggests most resolve without intervention. What the agent cannot see what no single team has connected yet is that a recent policy change altered customer behavior just enough to invalidate that assumption.

Nothing is "broken." The system is simply learning from yesterday's truth.

Richard sees the pattern and stops the meeting.

"This isn't a bug," he says. "This is learning without ownership."

From MLOps to L-Ops

The organization already has MLOps pipelines—models are versioned, deployments are controlled, and metrics are tracked. And yet, none of these answers the question Richard is asking. "Who owns the *behavior* of this system over time?"

MLOps manages artifacts. L-Ops manages outcomes.

That difference becomes the cornerstone of the next shift.

Richard introduces a new discipline not as a rebrand, but as a necessity—*Learning Operations (L-Ops)*—not a team or a tool but an operating model.

What a Learning Loop Actually Looks Like

Richard asks the value stream team to map how learning *should* happen, not ideally, theoretically, or operationally.

They draw a loop:

Signals ➤ Interpretation ➤ Adjustment ➤ Action ➤ Outcome ➤ Signals

And then they slow it down.

- Who decides which signals matter?
- Who interprets them?
- Who authorizes adjustment?
- Who validates outcomes?
- Who intervenes when learning moves in the wrong direction?

In the old model, these questions were spread across functions and time horizons. In the new model, they collapse into one accountable loop. The value stream team does not just ship features. It:

- Reviews behavior regularly
- Examines drift deliberately
- Decides when to retrain, when to constrain, and when to pause
- Documents why learning changes occur, not just that they occurred

Learning becomes intentional, not incidental.

Example: Weekly L-Ops Review for the Dispute Resolution Agent

Every Tuesday, the value stream team has an L-Ops review, as shown in Table 7-1.

Table 7-1. *Weekly L-Ops Review*

Metric	Current	Prior Month
Auto-resolution rate	71%	76%
Avg handling time	+9%	Baseline
Human override rate	18%	9%
Partial refund complaints	+34%	Baseline
Cost per case	+12%	Baseline

Monitoring tools show rising overrides in one region and lower satisfaction scores.

The team investigates recent policy changes and decides to:

- Pause autonomous handling for partial refund disputes
- Route those cases to human review
- Revert escalation threshold to prior version
- Launch retraining using new policy data
- Reevaluate after seven days

The rollback is executed through workflow controls in the orchestration platform within minutes.

Why Continuous Learning Changes Accountability

Richard notices an important shift in language. Teams stop asking, "Who deployed this?" They start asking, "Why did the system learn this?"

Blame gives way to inquiry. Not because people are nicer, but because the operating model supports it.

With L-Ops:

- Learning is expected
- Drift is observable
- Change is routine
- Intervention is normal

Failures are no longer surprises. They are feedback. And feedback, when owned, is manageable.

The Hidden Cost of Ignoring Learning Loops

Richard also sees what happens when learning loops are *not* formalized.

Teams rely on:

- Dashboards without context
- Alerts without interpretation
- Reviews without authority

Everyone is watching the system. No one is steering it.

This is where agentic AI becomes dangerous, not because it is autonomous, but because autonomy is unmanaged.

L-Ops does not slow innovation; it prevents silent decay.

The Boundary This Principle Creates

Before closing the session, Richard draws a clear line.

"If we don't own learning," he says, "we don't own outcomes."

Learning loops cannot be an afterthought. They cannot be owned by a platform team alone. They cannot live in governance reviews disconnected from execution. They must live where decisions are made. This principle does not replace value stream alignment. It activates it. And it sets up the next tension Richard knows is coming.

Because once learning is continuous and owned, a new conflict emerges: How do you balance experimentation with reliability?

That question leads directly to the next design principle.

Design Principle 3: Dual Operating Model

Once learning becomes continuous, tension becomes visible. Richard notices it almost immediately.

The value stream team is working better than before. Drift is identified earlier. Adjustments are more deliberate. Outcomes are improving. But a new friction has emerged—one that feels familiar, even inevitable.

Some people want to move faster; others want to slow everything down.

The same agent that benefits from experimentation also operates in production. The same learning loop that enables improvement introduces risk. The same autonomy that creates efficiency threatens stability.

The organization is being pulled in two directions at once.

The Collision of Two Types of Work

Richard names the conflict plainly. "There are two kinds of work happening here," he says. "And we're pretending they're the same."

On one side is exploration:

- Trying new signals
- Experimenting with thresholds
- Testing alternative behaviors
- Learning what *might* work

On the other side is execution:

- Running reliably
- Meeting service expectations
- Complying with controls
- Protecting customers and the business

Both are necessary and valid and both suffer when forced into a single operating rhythm.

The organization has made this mistake before—during agile transformations, cloud migrations, and data platform rollouts. But agentic AI sharpens the conflict because experimentation and execution now affect the same system, at the same time.

Richard realizes the problem is not speed. It is mode confusion.

Why One Operating Model Cannot Serve Both

The team reviews recent decisions.

An experiment improves resolution speed in a sandbox. It is promising but noisy. Some want to push it live quickly. Others insist on extended validation.

Neither side is wrong. The mistake is asking one group to satisfy both expectations simultaneously.

Exploration requires:

- Tolerance for failure
- Loose constraints
- Rapid iteration
- Incomplete information

Execution requires:

- Predictability
- Discipline
- Clear ownership
- Conservative change

When these demands collide, organizations default to compromise. And compromise is exactly what fails both sides.

Richard writes two words on the board:

Explore | Exploit

"Not phases," he says. "Modes."

Separating Without Isolating

The solution is not to choose one over the other. It is to run both deliberately.

Richard introduces the idea of a dual operating model—not as theory, but as structure.

One mode is optimized for exploration:

- Sandboxed environments
- Rapid experimentation
- Safe-to-fail assumptions
- Learning-focused metrics

The other is optimized for execution:

- Hardened pipelines
- Strict controls
- Outcome guarantees
- Risk-aware decisioning

The key is not separation alone; it is intentional connection.

Experiments do not drift casually into production. Production does not absorb change accidentally.

There is a clear transition point—explicit, reviewed, and understood.

Not a gate that blocks innovation. A handoff that protects it.

What Changes When the Model Is Explicit

Once the dual model is articulated, behavior changes quickly.

Exploratory work accelerates because it no longer pretends to be production ready. Teams test bolder ideas without fear of breaking live systems.

Execution stabilizes because it no longer absorbs experimental volatility. Change becomes deliberate, not constant.

Most importantly, conflict becomes constructive. People stop arguing about *whether* to move fast or slow. They start agreeing on which mode they are in.

Richard observes the shift in meetings. Questions change from: "Why did this break?" to "Did this belong in exploration or execution?"

That clarity alone prevents dozens of future failures.

Why Agentic AI Makes This Non-Negotiable

In traditional systems, exploration and execution were separated by time. Experiments happened before launch. Production stabilized afterward.

Agentic AI collapses that separation; learning continues after deployment, signals evolve continuously, and improvement never really ends.

Without a dual operating model, organizations oscillate between recklessness and paralysis. With it, they gain control without suffocation.

Richard summarizes it simply:

> *"Exploration discovers value. Execution protects it. Confusing the two destroys both."*

The Tension That Remains

Even with two operating modes, one challenge remains. Exploration wants freedom; execution demands safety.

Someone has to define boundaries. Someone has to decide:

- What experimentation is allowed
- Which constraints are non-negotiable
- Which risks are acceptable
- Where the system must never go

That responsibility cannot live entirely inside product teams or risk functions alone. It requires a new approach to governance, one that enables speed without sacrificing trust. That realization leads directly to the next design principle.

Design Principle 4: Centralized Governance, Decentralized Execution

The hardest conversations begin when everyone agrees on the problem.

Richard sees it clearly now. The organization understands value streams. It has accepted continuous learning. It has separated exploration from execution. And yet, a final tension remains—one that cannot be resolved by structure alone.

Who decides the limits? Not the roadmap, sprint plan, or the experiment backlog. The limits that matter are different:

- What an agent is allowed to decide
- How far learning can go before intervention
- Which behaviors are unacceptable, even if effective
- Where risk outweighs efficiency

Those decisions define trust. And trust, Richard knows, does not emerge organically at scale.

Why Traditional Governance Fails Agentic Systems

Governance in the organization has always been centralized. Committees review designs, controls approve changes, and audits validate compliance. This model works when systems are static.

Agentic AI is not.

By the time a governance body reviews a decision, the agent has already learned from new data, adapted behavior, and influenced outcomes. Oversight lags reality—not because people are slow, but because the model is mismatched.

Richard hears the frustration on both sides.

Risk teams feel overwhelmed, Product teams feel constrained, Engineering teams feel caught in between. Everyone is doing their job. The system still feels unsafe.

The False Choice Between Control and Speed

The debate inevitably polarizes.

Either:

- Tighten governance and slow everything down
- Decentralize decisions and accept more risk

Richard rejects the framing outright.

"Control and speed are not opposites," he says. "Opacity and clarity are." The problem is not where governance sits. The problem is how it is expressed.

Traditional governance tries to approve decisions. Agentic systems require governance that shapes the decision space itself.

From Gatekeepers to Guardrails

Richard introduces the metaphor that finally unlocks alignment.

"Governance should build the rails," he says, "not drive the trains."

The idea is simple but powerful. Central governance defines:

- Data boundaries
- Ethical constraints
- Risk thresholds
- Evaluation standards
- Escalation rules
- Auditability requirements

These are non-negotiable; they do not change from team to team, do not vary by experiment, and are stable, explicit, and enforced by design. Within those rails, teams execute freely.

They choose:

- How agents behave
- Which signals to use
- How to optimize outcomes
- When to adjust strategies

Execution is decentralized. Responsibility is not.

What Changes When Governance Is Embedded

Once governance shifts from approval to enablement, behavior changes quickly. Teams stop bypassing controls—because controls are no longer external. Risk reviews stop being reactive because signals are visible by default. Audits become easier—because decisions are traceable, not reconstructed.

Most importantly, governance scales, not by reviewing more decisions but by making unsafe decisions impossible.

Richard notices something subtle but important—governance stops being feared; it becomes infrastructure.

The Role of Centralization

Richard is careful to clarify what must remain centralized.

Not decision-making, but standards. The organization centralizes:

- Definitions of acceptable risk
- Ethical principles
- Evaluation frameworks
- Monitoring expectations
- Intervention triggers

These create consistency. They allow decentralized teams to move quickly without reinventing safety each time.

Decentralization without central standards leads to chaos. Centralization without local execution leads to paralysis. The balance matters.

Why This Principle Completes the Foundation

With this principle in place, the operating model becomes coherent. Value stream teams own outcomes and learning loops are continuous and intentional. Exploration and execution coexist without conflict. Governance enables speed without sacrificing trust.

Each principle reinforces the others. Remove anyone, and the system weakens. Richard pauses at the end of the session and looks around the room.

"We're not giving up control," he says. "We're redesigning it."

The organization is no longer trying to supervise intelligence from outside.

It is beginning to embed responsibility into how intelligence operates. That realization marks the end of one phase of the journey and the beginning of another. Because once structure and governance are addressed, a deeper question emerges: "Do we have the right people, skills, and roles to make this work?"

New Roles and Skills for the AI Era

Once governance shifts from gatekeeping to guardrails, Richard notices something else surface almost immediately.

The questions in meetings change.

- Not what should the agent do, but who is actually equipped to decide.
- Not how do we deploy, but who understands the consequences of behavior over time.
- Not who approved this, but who is accountable for learning when it goes wrong.

The operating model now makes sense. The people model does not.

The organization has redesigned how work flows—but not how responsibility is held. And responsibility, Richard knows, always reveals gaps in skills long before it reveals gaps in technology.

Why Old Roles Start to Bend

No one's job description explicitly says, "own AI behavior."

Product managers are trained to define features and prioritize roadmaps. Engineers are trained to build systems and ensure reliability. Data scientists are trained to optimize models and accuracy. Risk teams are trained to prevent failure.

Agentic systems require all of these simultaneously. The result is quiet overload.

Product managers are asked to make ethical trade-offs that they were never trained for. Engineers are asked to reason about probabilistic outcomes instead of deterministic logic. Data scientists are expected to explain behavior to regulators and customers. Risk teams are pulled into day-to-day execution instead of strategic oversight.

No one is failing. The roles are simply misaligned with the work.

The Shift from Tasks to Stewardship

Richard reframes the issue in a leadership session. "We're not missing talent," he says. "We're missing *stewards*."

Agentic AI does not need more hands. It needs clearer ownership of:

- Intent
- Behavior
- Learning
- Boundaries

These responsibilities do not map cleanly onto traditional roles. So the organization stops pretending they do.

The Emergence of New Roles

Rather than inventing titles for novelty's sake, Richard insists on naming work that already exists but is currently invisible.

AI Product Manager

This role looks like a product manager on paper, but the center of gravity shifts.

The AI product managers do not just ship features. They define acceptable behavior. They balance accuracy, speed, trust, and risk. They decide when automation should defer to humans. They are accountable not just for outcomes—but for *how* those outcomes are achieved.

AI Behavior Designer

As agents interact more directly with customers and employees, behavior becomes part of the product:

- Tone
- Escalation
- Uncertainty
- Deference

These are not technical details. They shape trust. The AI behavior designer focuses on how agents behave under ambiguity—not when everything goes right, but when it doesn't.

Model Evaluator/Learning Steward

Someone must own the question: "Is the system learning the right things?"

This role sits at the intersection of data science, risk, and operations. They monitor drift, evaluate unintended consequences, and initiate intervention. They are not optimizing models. They are safeguarding behavior over time.

AI Systems Architect

Traditional architects optimize for scalability and resilience. AI systems architects add other dimensions:

- Decision flow
- Autonomy boundaries
- Feedback loops
- Escalation paths

They design systems where intelligence is distributed—but controlled.

Embedded Risk and Ethics Partners

Risk and ethics no longer live downstream. They embed directly into value streams—not to approve work, but to shape it early. They help define rails, not review wreckage.

Skills Matter More Than Titles

Richard is careful not to over-index on org charts. The real shift is in skills.

Teams now need:

- Probabilistic thinking instead of binary logic
- Comfort with uncertainty

- The ability to reason about systems, not components
- Fluency in trade-offs rather than rules
- Literacy in how learning systems behave over time

These skills cut across roles. Training, hiring, and career paths begin to reflect that reality.

The Cultural Shift That Follows

As roles clarify, something unexpected happens.

Fear decreases. People stop worrying about being blamed for things they don't control. They stop making defensive decisions. They start asking better questions.

The organization moves from "Who approved this?" to "Who owns this learning?".

That single shift changes everything. Richard sees it clearly now. You cannot run agentic systems with roles designed for static software. Not because people are incapable, but because the work has changed.

The Door This Opens

With clearer roles and emerging skills, the organization becomes more confident but also more fragmented. Different teams are solving similar problems. Patterns are emerging but unevenly. Hard won lessons risk being lost.

Richard recognizes the next challenge.

If value streams create focus and roles create accountability, something else is needed to create coherence across the organization. That realization leads to the next structural layer.

The AI Guild Model

Clear roles solve one problem. They do not solve another fragmentation.

Richard sees it emerging as soon as value stream teams and new responsibilities take hold. Each team is doing better work. Learning loops are owned. Decisions are clearer. Outcomes improve. But the work is uneven.

One team develops a strong approach to evaluating drifts. Another invents its own way of testing agent behavior under stress. A third refines escalation logic through painful trial and error. The organization is learning, but it is learning in pockets.

That pattern is familiar. Richard has seen it before with agile teams, data science groups, and platform migrations. Without an intentional mechanism for sharing craft, organizations trade local excellence for global inconsistency.

This is where the *AI Guild Model* enters, not as a hierarchy but as connective tissue.

Why Guilds Exist

Guilds are not delivery teams. They do not own roadmaps. They do not approve of work. They exist to preserve mastery across a decentralized system. Richard describes it simply:

"Value stream teams deliver outcomes. Guilds ensure we don't forget how."

The distinction matters.

As AI systems evolve, knowledge decays quickly. Techniques that work in one context are rediscovered or relearned in another. Mistakes repeat. Patterns fragment.

Guilds counteract that drift.

How the Guild Model Works

Each guild is organized around a shared discipline, not a product. The organization forms a small number of AI-focused guilds:

- **AI engineering guild:** Deployment patterns, reliability practices, tooling standards
- **AI product and behavior guild:** Behavior design, escalation patterns, trust heuristics
- **Learning and evaluation guild:** Drift detection, testing frameworks, red-teaming practices
- **Data and signals guild:** Data contracts, observability, signal quality
- **Governance and risk guild:** Guardrails, auditability, ethical interpretation

Membership is lightweight.

People remain embedded in value stream teams. They meet periodically with peers who do similar work elsewhere. They share patterns, failures, and lessons learned.

What Guilds Change in Practice

The impact is subtle but compounding.

Teams stop reinventing evaluation frameworks. Common language emerges around acceptable behavior. New hires ramp faster because patterns are documented and socialized. Risk teams see consistency instead of surprise.

Most importantly, guilds create horizontal alignment without reintroducing central control.

Richard notices that some of the best insights no longer come from leadership meetings. They come from guild conversations.

People talk openly about what failed—not to assign blame, but to improve discipline. That culture is difficult to mandate.

Guilds make it possible.

The Boundary That Keeps Guilds Healthy

Richard is explicit about one rule. "Guilds do not own decisions." They influence, advise, and educate.

But accountability remains with value stream teams. This prevents two common failure modes:

- Guilds turning into shadow hierarchies
- Guilds becoming bottlenecks

When this boundary is respected, guilds scale naturally. When it is not, they collapse under their own weight.

Why Guilds Fail in the Real World

Richard had also seen guilds fail before—not through conflict, but through irrelevance.

They begin with energy, strong participation, and shared enthusiasm. Over time, calendars tighten. Senior practitioners stop attending. Meetings become optional forums with little consequence. Discussions repeat familiar topics. Outputs disappear into notes no one reads.

The guild still exists on paper but no longer shapes how work gets done.

Richard considered this the most common guild failure mode. "People don't abandon guilds because they dislike learning," he said. "They abandon guilds when learning stops changing anything."

To prevent this, he introduced the three disciplines discussed next.

Executive Sponsorship Without Control

Each guild had a senior sponsor who reinforced its importance, removed blockers, and ensured outputs were visible. Sponsors did not run guilds, but they signaled that participation mattered.

Rotating Guild Leadership

Guild leads rotated periodically across respected practitioners. This prevented burnout, avoided gatekeeping, and kept fresh thinking flowing into the discipline.

Tangible Outputs Linked to Standards

Guild insights had to translate into something real:

- Updated engineering standards
- Approved evaluation playbooks
- Reusable templates
- Governance controls
- Onboarding guides
- Platform backlog priorities

When guild outputs shaped the operating model, attendance stayed valuable.

Note If a guild does not improve how work happens, it becomes a meeting.

Why Guilds Matter More for AI Than Software

Software patterns change slowly. AI patterns change constantly. Model behavior shifts. Regulatory expectations evolve. New failure modes emerge.

Without a mechanism for collective learning, organizations fall behind their own systems.

Guilds ensure that learning keeps pace with intelligence. They do not replace leadership. They amplify it.

The Shape That Emerges

As value streams, learning loops, roles, and guilds settle into place, Richard steps back and observes something important. There is no single org chart anymore. There is a system that supports:

- Vertical ownership through value streams
- Horizontal coherence through guilds
- Central standards through governance
- Local execution through teams

The shape is intentional. And yet, it is not the only shape possible. Different organizations will need different forms. That realization leads naturally to the next question: "What are the viable organizational archetypes for AI teams and when should each be used?"

Organizational Archetypes for AI Teams

At this point, Richard is careful not to present what they've built as the model. It is a model that fits their context, maturity, and risk profile.

That distinction matters.

Over the years, Richard has watched organizations copy structures from others they admire, only to discover later that the structure was never the real source of success. Context was.

AI makes that mistake more expensive.

So instead of prescribing a single design, Richard frames what they've learned as a set of organizational archetype patterns that appear repeatedly, each with strengths, limitations, and a narrow range of contexts where they work well.

The question for leaders is not which is best, but which is appropriate now.

Archetype 1: Centralized AI Hub

In this model, AI capability lives primarily in a central team.

Data scientists, AI engineers, platform specialists, and governance experts are grouped together. Business units consume AI as a service through APIs, shared platforms, or internal consulting. This archetype excels in the early stages of AI adoption. It creates:

- Consistency
- Strong governance
- Efficient use of scarce expertise

But it struggles as agentic systems scale.

The distance between those building intelligence and those living with its consequences grows. Feedback slows. Context is lost. The hub becomes a bottleneck, or worse, a shield that isolates AI from reality.

Richard has seen this model succeed for experimentation. He has rarely seen it succeed for ownership.

Archetype 2: Federated AI Value Streams

This is the model Richard's organization is moving toward.

AI capabilities sit inside product-aligned value stream teams. Each team owns outcomes end to end, supported by shared platforms and guilds. Centralization shifts from people to standards. This archetype balances autonomy with coherence. It allows:

- Fast learning
- Local accountability
- Continuous improvement

But it requires discipline. Without strong governance rails and active guilds, it devolves into fragmentation. With them, it scales remarkably well.

Richard considers this the most robust model for organizations running multiple AI-powered products at scale.

Archetype 3: AI Platform-Centric Organization

Some organizations go further.

They invest heavily in internal AI platforms—shared tooling, reusable agents, evaluation frameworks, and safety infrastructure. Product teams build on top of these platforms rather than reinventing intelligence each time.

This archetype works best when:

- AI usage is widespread
- Engineering maturity is high
- Platform investment is sustained

Its strength is leverage. Its risk is detachment.

When platforms drift too far from real-world use cases, they optimize elegance over impact. Without strong product alignment, they become impressive, but underutilized.

Archetype 4: Agent First Organization

This archetype appears most often in startups.

Teams organize around agents rather than functions. Small groups own cohorts of agents that perform specific jobs. Boundaries are fluid. Roles overlap.

This model enables extraordinary speed. It also carries extraordinary risks. Without formal governance, learning loops can outrun oversight. Without clear ownership, scaling magnifies fragility.

Richard views this archetype as powerful but unstable. It demands exceptional leadership maturity to sustain beyond early growth.

Choosing an Archetype Is a Leadership Decision

Richard emphasizes that these archetypes are not evolutionary stages in a fixed sequence. They are choices. Each reflects:

- Risk tolerance
- Regulatory environment
- Product complexity
- Leadership maturity

The mistake is copying what worked elsewhere without understanding why. AI does not reward structural imitation. It rewards an intentional design.

The Pattern Beneath the Patterns

Despite their differences, Richard notices something common across successful organizations, regardless of archetype.

They all:

- Align ownership to outcomes
- Make learning visible
- Separate exploration from execution
- Embed governance into how work is done

Structure varies. Principles do not.

That realization becomes the foundation for the final part of the chapter. Because understanding archetypes is useful—but leadership requires action.

Richard now turns from patterns to practice.

Richard's Transformation Blueprint

Richard is careful not to present what comes next as a grand reorganization.

He has learned that large structural resets create the illusion of progress while postponing the real work. Org charts change quickly. Behavior does not.

Instead, he frames the transformation as a sequence of deliberate moves—each small enough to execute, but together powerful enough to reshape how the organization works with intelligence.

He calls it a blueprint, not because it is rigid, but because it provides orientation when complexity rises.

Step 1: Make Outcomes Explicit

The first change is deceptively simple.

Every AI-powered product is required to articulate a clear outcome—not a feature set, not a roadmap, but a measurable change in the world. For the dispute-resolution agent, the outcome is not automation. It is fair, fast, and consistent resolution.

This forces teams to align discussions around behavior, not output. It also makes trade-offs visible early. When outcomes are explicit, accountability has somewhere to land.

Step 2: Assign End-to-End Ownership

Next, Richard formalizes what had previously been implicit. Each AI-powered product has a single value stream owner—someone accountable for how the system behaves in production over time.

Not a coordinator. Not a sponsor.

An owner.

This does not collapse all responsibility into one person. It clarifies who ensures that responsibility is not fragmented.

When questions arise about drift, escalation, or unintended consequences, everyone knows where to start.

Step 3: Institutionalize Learning Reviews

Richard resists the temptation to turn learning into dashboards alone. Instead, he introduces a new cadence. Value stream teams hold regular learning reviews focused not on delivery metrics, but on:

- What the system learned
- Why it learned it
- Whether that learning is still valid
- What boundaries need adjustment

These reviews include product, data, engineering, and risk voices by default.

Learning becomes visible. Intervention becomes normal.

Step 4: Explicitly Separate Exploration from Execution

The dual operating model is no longer theoretical. Exploration environments are clearly labeled. Execution environments are protected.

Transitions between the two are intentional, reviewed, and documented—not bureaucratic, but deliberate. This removes ambiguity. Teams know when speed is expected. They know when caution is required.

Conflict decreases because expectations are explicit.

Step 5: Build Governance into the System

Richard works with risk and compliance leaders to shift governance upstream. Rather than reviewing every decision, they define:

- Non-negotiable constraints
- Escalation thresholds
- Monitoring expectations
- Audit trails

These are embedded into platforms and workflows. Governance becomes an infrastructure, not an event.

Step 6: Name the New Work

New responsibilities are no longer hidden inside old roles.

AI product stewardship, behavior design, learning oversight, and system boundary design are formally recognized.

This does not inflate the org chart. It legitimizes work that already exists.

People stop improvising accountability.

Step 7: Connect Teams Through Guilds

Guilds are formalized with light structure.

- They have charters.
- They have shared artifacts.
- They have regular forums.

They do not approve work. They accelerate learning. This prevents local excellence from becoming organizational amnesia.

Step 8: Choose the Right Archetype, Consciously

Finally, Richard helps leadership name the organizational archetype they are committing to, at least for now. This removes ambiguity. People stop debating structure implicitly and start evaluating it explicitly as conditions change.

The organization gains a shared mental model.

What the Blueprint Does and Does Not Do

Richard is clear about one thing. This blueprint does not guarantee success. It does something more important. It makes failure visible early, understandable, and recoverable.

Why the Blueprint Is Cyclical, Not Linear

Richard was equally clear about something leaders rarely admit: organizations regress.

A leadership change can reintroduce silos. A major compliance event can recentralize decision-making. A failed AI launch can trigger fear and overcorrection. Cost pressure can push teams back into short-term delivery mode. High performers leaving can weaken guilds and ownership models.

Progress does not move in a straight line.

Structures that took years to improve can drift backward in a quarter if incentives, leadership attention, or trust changes.

That is why Richard never described the blueprint as a one-time transformation plan. It was a recurring management discipline.

These steps need to be revisited repeatedly:

- Outcomes reclarified as priorities shift
- Ownership reset as teams evolve
- Learning reviews strengthened when drift appears
- Governance recalibrated after incidents
- Guilds renewed when participation fades
- Organizational archetypes reconsidered as scale changes

The blueprint was less like a construction project and more like organizational maintenance.

> *"You do not arrive at an AI operating model. You keep returning to it."*

That, he knows, is the real measure of readiness in an AI-driven organization.

Richard's Realization

The change does not announce itself.

There is no single moment where Richard can point and say, this is when it worked. No celebratory email. No dramatic dashboard spike. No applause.

Instead, realization arrives quietly.

It comes during a routine review—one that would once have been tense, defensive, and overly detailed. The dispute resolution agent has handled a new category of cases. Outcomes are mixed.

A few decisions are reversed. Learning thresholds are adjusted. No one argues about ownership. No one debates whether the issue belongs to product, data, engineering, or risk.

The conversation simply moves forward. Richard notices the absence of friction before he notices anything else.

Not because problems have disappeared but because they surface earlier, with context, and with people who are empowered to act. Decisions feel less political. Trade-offs feel more explicit. Learning feels intentional.

For the first time, the organization is not reacting to intelligence; it is working with it. He thinks back to the misfiring orchestra.

The problem had never been talent; it had never been effort or even the technology. It had been coordination without coherence. AI agents had simply revealed it faster than anything before.

What surprises Richard most is not that the new operating model works. It is how quickly people adapt once ambiguity is removed. When responsibility is clear, fear recedes. When learning is expected, mistakes lose their stigma. When governance is embedded, trust increases rather than erodes.

The organization feels calmer—not slower, but steadier.

Richard walks past the operations floor later that evening. He watches engineers reviewing agent behavior alongside product managers and risk partners. Not in a crisis. Not under pressure. Just as part of normal work.

This is what had been missing all along.

Not better models or smarter tools, but an organization designed for systems that change.

He realizes something that had not been obvious when the journey began:

> *"Leadership in the age of AI is no longer about directing intelligence. It is about designing the conditions under which intelligence can operate safely, responsibly, and at scale."*

That shift is subtle and irreversible. Richard also understands what this transformation is *not*. It is *not* about eliminating human judgment, removing accountability, or handing control to machines.

It is about recognizing that once intelligence becomes embedded in work, leadership must move upstream—from decisions to design. From approvals to boundaries and from control to coherence.

As he leaves the building, Richard knows this chapter is not the end of the story; it is a new baseline.

The organization will evolve again. New agents will appear. New risks will emerge. Some assumptions will break. But the organization now has something it did not have before. A way to adapt deliberately.

That, more than any technology choice, feels like the real competitive advantage.

Reflection Questions

- Where in your organization do AI systems already operate end to end—while ownership remains fragmented?
- Who is accountable for how AI-powered products *behave over time*, not just how they are built or deployed?
- Are your teams organized around functional excellence—or around outcomes that customers actually experience?
- Do you treat learning, drift, and adaptation as incidents—or as normal operating conditions?
- Where are exploration and execution unintentionally competing inside the same teams?
- Does your governance model approve decisions after the fact—or shape which decisions are possible in the first place?
- Which responsibilities introduced by agentic systems exist today but are still unnamed, unowned, or informal?
- If an AI agent changed its behavior tomorrow in a way that mattered, would your organization know *where* to intervene—and *who* should act?

Note AI readiness is no longer a question of tools or talent. It is a question of organizational design.

CHAPTER 8

Governance As an Innovation Engine, Not a Bottleneck

Governance does not appear at the beginning of transformation. It arrives later quietly and is usually uninvited.

Richard realizes this not during the redesign sessions, nor during the early wins that follow. It shows up when things start working. When AI-powered products begin to scale. When learning accelerates. When decisions move faster than comfort allows.

That is when governance finally shows up, not as a concept, but as friction.

The dispute resolution agent is performing well. Better than expected, in fact. Resolution times are down. Customer complaints are trending lower. Teams trust the system enough to let it handle increasingly complex cases.

And that is precisely when the questions begin.

Risk leaders ask how escalation thresholds are evolving. Compliance asks how decisions are being logged and explained. Audit asks what happens when learning changes outcomes months after deployment.

S. Shivam, *The Leadership Algorithm*, https://doi.org/10.1007/979-8-8688-2897-3_8

None of these questions are unreasonable. What surprises Richard is how late they arrive. Governance did not slow the early experimentation, did not challenge the pilots, and did not question the first deployments.

It surfaces at the moment of confidence; intelligence stops being interesting and starts being consequential.

Richard sits in a meeting where no one is arguing, yet the tension is unmistakable. Product wants to expand the agent's scope. Engineering wants to stabilize the pipeline. Risk wants stronger constraints. Compliance wants clearer traceability.

Everyone agrees on the goal. No one agrees on the path. This is not resistance to innovation. It is uncertainty about control.

Governance has always existed in the organization.

Policies were written, reviews were conducted, and audits were passed.

But those mechanisms were built for systems that behaved predictably once approved—for software that stayed still unless humans intervened.

Agentic AI does not behave that way. It learns; it adapts. It accumulates small decisions into meaningful outcomes.

By the time governance looks at the system, the system has already moved.

Richard recognizes the pattern immediately. Governance is not failing because people are cautious. It is failing because it is being asked to govern something it was never designed to see in motion.

In earlier eras, governance functioned as a checkpoint. You paused progress, reviewed risk, approved change, and moved on. That model assumes that risk can be assessed at moments in time. Agentic systems stretch risk across time. They turn governance from an event into a continuous concern. And yet, the organization still treats it as episodic.

That mismatch is where innovation begins to slow—not because governance says "no," but because it does not know how to say "yes" with confidence.

Richard reframes the situation in his own mind. This is not a failure of governance intent. It is a failure of governance design.

The organization has redesigned teams, ownership, learning loops, and operating models—but governance has remained largely unchanged. It still sits outside the system, observing outcomes instead of shaping behavior.

As long as that remains true, two things will happen simultaneously:

- Teams will feel constrained just as they gain momentum
- Risk leaders will feel exposed just as intelligence scales

Neither side will trust the system fully.

That realization lands with clarity. Governance is not the last problem to solve in AI transformation. It is the next one. Not because governance must control intelligence—but because intelligence must be designed to operate within boundaries that make trust possible.

Richard writes a single sentence at the top of a fresh page in his notebook:

"If governance only appears at scale, it is already too late."

The rest of the thought flows from that insight. Because the question is no longer whether governance matters. It is whether governance can evolve fast enough to become an engine for innovation rather than the place innovation goes to slow down.

Why Traditional Governance Breaks with AI

Richard has seen governance work well before. It worked when systems were static, when change was deliberate, and when risk could be evaluated at specific moments in time.

None of those conditions hold anymore.

Traditional governance assumes that a system can be understood, approved, and then trusted to behave within known boundaries. Once approved, the system is monitored primarily for availability and compliance—not for evolution.

AI breaks that assumption quietly and completely.

The system Richard is now responsible for does not stay the same after approval. It changes with data. It changes with usage. It changes with context. It does not wait for a review cycle to adapt its behavior.

Governance, meanwhile, still waits.

The Checkpoint Fallacy

Most governance models are built around checkpoints—design reviews, model approvals, release signoffs, and periodic audits.

Each checkpoint asks the same underlying question: is this safe to proceed right now?

For traditional software, that question is meaningful. The system will behave tomorrow much as it behaves today unless a human explicitly changes it.

Agentic AI invalidates that premise.

The most important risks do not appear at checkpoints. They emerge between them slowly, cumulatively, and often invisibly.

By the time a governance body reconvenes, the system has already learned from hundreds of new interactions, internalized new patterns, and influenced outcomes in ways that no single decision captured.

Richard realizes that governance is not failing because it is too slow. It is failing because it is looking in the wrong places.

Static Risk Models in a Dynamic World

Traditional governance also assumes that risk can be cataloged. Known risks are documented, mitigations are defined, and residual risk is accepted.

This works when risk is stable. Agentic AI introduces emergent risk.

No one explicitly programs the system to behave incorrectly. Instead, small, reasonable adjustments interact in unexpected ways. Learning compounds. Context shifts. Edge cases accumulate.

The system does not violate rules.

It drifts toward outcomes that no one intended, but everyone technically allowed.

Richard notices how uncomfortable this makes risk teams.

They are asked to sign off on systems whose future behavior cannot be fully enumerated. They are asked to approve learning without knowing exactly what will be learned.

This is not a failure of diligence. It is a mismatch between governance tools and system behavior.

The answer is not to freeze learning. It is to govern learning itself. That requires moving beyond static approvals toward continuous assurance.

Leading organizations already apply disciplines from MLOps to make adaptive systems governable. Model drift detection monitors when behavior, inputs, or outcomes begin to diverge from expected ranges. Champion-challenger frameworks compare incumbent models against newer candidates in controlled conditions before full deployment. Automated retraining pipelines refresh models using approved data, tested workflows, and documented controls rather than ad hoc updates. Rollback mechanisms allow rapid reversion to prior stable versions when performance, fairness, or risk thresholds are breached.

In this model, drift is not a failure. Silence is.

Governance becomes less about predicting every possible future state and more about ensuring that when systems change, the organization can detect it, understand it, and respond with speed and discipline.

The Illusion of Oversight

As governance struggles to keep pace, organizations often respond by adding more oversight. More reviews, more documentation, and more approvals.

The intent is sound; the effect is not.

Oversight increases lag without increasing understanding. Teams spend more time justifying decisions than examining outcomes. Governance becomes performative and focused on artifacts rather than behavior.

Meanwhile, the system continues to learn.

Richard sees the danger clearly.

When governance focuses on process compliance instead of behavioral understanding, it creates a false sense of safety.

The organization feels controlled; the system is not.

Why Committees Cannot Govern Learning

Committees are excellent at evaluating decisions; they are terrible at governing adaptation.

Learning systems do not change in discrete steps. They evolve continuously. Governing them requires sustained attention, contextual awareness, and the ability to intervene quickly when signals emerge.

No committee can meet often enough—or deeply enough—to provide that kind of stewardship.

This is not a criticism of governance bodies. It is a recognition of their design limits.

Richard writes another note to himself:

"You cannot govern learning systems by reviewing decisions.

You must govern the conditions under which decisions are made."

That sentence becomes the hinge for everything that follows.

The Cost of Delay

The final failure mode Richard observes is subtle. As governance struggles to keep up, teams begin to self-regulate.

They slow down experimentation. They avoid ambitious use cases. They choose safer, less impactful paths.

Innovation does not stop. It shrinks.

This is how governance becomes a bottleneck—not by saying "no," but by making "yes" too costly.

And once that pattern sets in, it is difficult to reverse.

Richard steps back from the details and sees the full picture.

Traditional governance is not wrong. It is simply optimized for a world where intelligence is static, change is episodic, and risk can be assessed at moments in time.

AI operates in none of those modes.

If governance is to remain relevant—if it is to enable rather than constrain—it must be redesigned as deliberately as the organization itself was redesigned in the previous chapter.

That redesign begins with a shift in how governance is understood. Not as control, but as infrastructure.

Governance As Infrastructure, Not Control

Richard realizes that the word governance itself has become part of the problem.

In most organizations, governance is spoken about in the same breath as restriction. It is something you "go through," something that "slows things down," something that appears when risk is already high.

That framing is backward. Governance does not fail because it is too strict. It fails because it is too late.

Once Richard sees this, the solution becomes less about tightening rules and more about redesigning where governance lives.

The Difference Between Control and Constraint

Control operates after intent is expressed. You make a decision. Someone reviews it. Someone approves or rejects it.

Constraint operates before intent is expressed. It defines what is possible in the first place.

Richard explains the difference to his leadership team using a simple analogy:

Traffic lights control behavior. Road design constrains it. One requires constant enforcement. The other shapes outcomes by default. Traditional governance behaves like traffic lights. AI-native governance must behave like road design.

Why Infrastructure Scales When Oversight Does Not

Oversight depends on attention. Infrastructure depends on design. As AI systems scale, the number of decisions they make grows exponentially. No governance function—no matter how well staffed—can review each one meaningfully.

Infrastructure solves this by encoding governance into the system itself.

Richard begins to shift conversations away from approval workflows and toward design questions:

- Which decisions should be impossible for an agent to make?
- Which actions must always trigger escalation?
- What data should never be used for learning?
- Which outcomes require human confirmation, regardless of confidence?

These are not review questions. They are architectural choices. Once made, they do not need to be re-decided every week.

From Policies to Rails

The organization already has policies. They describe what should and should not happen. But policies require interpretation. Interpretation requires judgment. Judgments vary.

Infrastructure removes ambiguity.

Richard works with governance, platform, and product leaders to translate abstract policies into concrete rails:

- Risk thresholds are embedded into pipelines, not documented in slides.
- Audit trails are automatic, not manually assembled.
- Escalation paths are enforced by system logic, not social norms.
- Learning boundaries are coded, not implied.

Governance stops being something teams comply with. It becomes something systems enforce.

Why This Accelerates Innovation

The fear, of course, is that constraints will slow teams down. Richard sees the opposite. Once constraints are explicit and embedded:

- Teams stop second-guessing what is allowed
- Experimentation becomes safer, not riskier
- Innovation moves faster because uncertainty is reduced

Teams do not need permission to move quickly. They need confidence that they are not creating invisible risks. Infrastructure provides that confidence.

The Shift in Governance Identity

This transformation also changes how governance teams see themselves. They are no longer:

- Reviewers of work
- Approvers of change
- Blockers of progress

They become:

- Designers of safe decision spaces
- Builders of trust mechanisms
- Stewards of long-term integrity

This is not a loss of authority. It is an increase in leverage.

Richard notices that once governance is expressed as infrastructure, conversations improve.

Instead of asking, "Can we do this?" Teams ask, "How do we design this safely?"

That single shift changes the tone of innovation entirely.

What This Makes Possible Next

Governance as infrastructure establishes the foundation. But agentic AI introduces further complications. These systems do not just operate within constraints. They adapt within them.

They learn, drift, and explore edge cases.

Governing static behavior is one challenge. Governing adaptive behavior is another entirely.

Richard knows the next question they must answer: "How do you govern systems that can act, learn, and change—without freezing them in place?"

Governing Agentic Systems Without Slowing Them Down

Once governance becomes infrastructure, new anxiety emerges. Not from governance teams—but from leaders.

If agents can act within constraints, learn continuously, and adapt faster than humans can review decisions, how do you ensure that speed does not quietly outrun control?

Richard understands the concern. It is valid.

Agentic systems are not dangerous because they are fast.

They are dangerous when speed and responsibility drift apart.

The goal of governance, he reminds the group, is not to slow systems down. It is to ensure that when systems move quickly, they do so intentionally.

When One AI Becomes Many

Richard notices the next governance problem arrives before most leaders are ready for it. The organization is no longer managing one intelligent system. It is managing many.

An agent drafts a recommendation. Another validates it. A third executes a transaction. A fourth learns from the outcome.

No single model owns the final decision, yet together they create it.

So who is accountable when Agent B acts on Agent A's flawed output?

Who owns the error when each component performed exactly as designed, but the chain produced the wrong result?

Traditional governance was built for standalone systems. Multi-agent environments require governance of interactions—handoffs, dependencies, and escalation rights.

Leadership must define:

- Decision rights across agents
- Human override points
- Traceability across agent chains
- Liability ownership for composite outcomes
- Approval thresholds for autonomous actions

The risk is no longer model failure. It is coordination failure.

When Action Replaces Recommendation

Most governance frameworks were designed for advisory systems.

Models predicted.

Dashboards informed.

Humans decided.

Agentic systems invert that flow. They decide first. Humans intervene only when thresholds are crossed.

This single shift changes everything.

Governance is no longer about approving decisions. It is about defining *which decisions require approval at all.*

Richard reframes governance questions accordingly:

- Which actions can an agent take autonomously?
- Which actions require confirmation?
- Which actions are prohibited entirely?

These questions are answered *once,* through design, not repeatedly, through review. Autonomy becomes bounded, not binary.

Escalation As a First-Class Design Concept

In traditional systems, escalation is an exception. In agentic systems, escalation is a feature.

Richard insists that escalation logic be treated with the same care as core functionality. Not as an afterthought. Not as a fallback.

Escalation answers three critical questions:

- When should the system ask for help?
- Who should it ask?
- What context should be provided automatically?

Well-designed escalation does not slow systems down. It prevents silent failure.

Richard notices a pattern across teams: when escalation rules are explicit, autonomy increases—not decreases. Teams trust the system because they know it will ask for help before crossing critical boundaries.

Governing Learning Without Freezing It

The hardest governance problem is learning. Not whether systems should learn, but how much, how fast, and from what.

Richard draws a simple distinction. Learning is not inherently good. Unbounded learning is dangerous.

Agentic systems must be governed not just on actions, but on how learning influences future actions.

He works with teams to define:

- Which signals are safe for learning
- Which signals require validation
- Which signals should never be learned from
- When learning must pause or roll back

Learning becomes reversible. Not perfect, but recoverable. That reversibility allows innovation to continue safely.

Drift Is Not Failure, Silence Is

Richard challenges one of the most persistent misconceptions. Drift is not evidence that a system is out of control. It is evidence that the environment is changing. The governance failure occurs when drift is:

- Invisible
- Unexplained
- Ignored

Effective governance makes drift observable and discussable.

Teams stop asking, "Why did this break?" They start asking, "What changed—and why?"

That shift turns governance into sense-making rather than enforcement.

Red Teaming As an Ongoing Capability

In traditional governance, testing happens before launch. Agentic systems demand more. Richard reframes red teaming not as a security exercise, but as a learning discipline.

Teams simulate:

- Adversarial behavior
- Unexpected user strategies
- Edge cases the system has not yet seen
- Combinations of signals that could push behavior toward unsafe outcomes

This is not about catching mistakes. It is about discovering blind spots early. Red teaming becomes routine, not reactive. And because it is embedded into the operating model, it accelerates confidence rather than eroding it.

Why Speed Improves When Governance Is Right

Richard notices a paradox. As governance becomes more explicit, teams move faster. Not because constraints are looser but because uncertainty is reduced.

People no longer waste time debating what is allowed. They no longer self-censor innovation out of fear. They know where the edges are. That clarity is what enables speed.

Governance does not slow agentic systems down. Poorly designed governance does.

The Boundary Between Trust and Control

Richard captures the principle succinctly: "Trust is not the absence of control. It is control designed in advance."

Agentic systems force leaders to confront this truth. You cannot approve every decision. You cannot predict every outcome. But you can design systems so that:

- Unsafe behavior is constrained
- Learning is bounded
- Intervention is fast
- Recovery is possible

That is what governance looks like when it enables innovation.

What This Makes Visible Next

With governance now shaping autonomy, escalation, and learning, a final question emerges: "Where does this governance capability live?"

Not as a function, not as a committee, but as an organizational responsibility.

Richard knows that answering that question requires one last integration step—bringing governance into the fabric of teams, platforms, and leadership itself.

Where Governance Lives in the Organization

Once governance is reframed as infrastructure, a practical question surfaces quickly. If governance is not a gate, not a committee, and not an overlay, where does it actually live?

Richard knows this question is often answered poorly.

Some organizations push governance entirely into a central risk or compliance function, distancing it from day-to-day work. Others distribute it so widely that no one truly owns it. Both approaches fail in different ways.

The answer, Richard realizes, lies in separation of responsibility without separation of presence.

Centralize Standards, Decentralize Execution

Governance must be centralized where consistency matters.

Standards for:

- Acceptable risk
- Ethical boundaries
- Escalation thresholds
- Auditability
- Evaluation criteria

These cannot vary by team without creating confusion and exposure. They define the organization's posture—not a product's preference.

But governance must be decentralized where context matters.

Execution, interpretation, and day-to-day judgment belong inside value stream teams. These teams live with the consequences of AI behavior. They see how decisions play out in real workflows. They are closest to signals that matter.

Richard makes the distinction explicit: "Governance defines the rules of the road. Teams decide how to drive—within them."

Governance Is a Capability, Not a Department

Richard resists the instinct to "stand up" a new governance team.

Departments optimize for throughput; governance optimizes for trust. Those are not the same thing.

Instead, governance becomes a capability embedded across the organization, supported by a small central group that:

- Defines standards
- Maintains shared frameworks
- Evolves constraints as conditions change
- Monitors systemic risk

This group does not approve work. It enables safe work.

Their success is measured not by how many decisions they review but by how rarely the teams are surprised.

The Role of Value Stream Teams

Value stream teams do not outsource governance. They operationalize it. Each team is responsible for:

- Applying governance constraints correctly
- Monitoring agent behavior
- Triggering escalation when boundaries are approached
- Documenting learning decisions and reversals

Governance is not something that happens *to* teams. It is something that teams practice. This shift changes behavior.

Teams stop viewing governance as external pressure. They begin to treat it as part of product quality.

Platforms As Governance Carriers

Richard sees platforms emerge as a critical leverage point. Governance scales best when it is carried by platforms rather than people.

For years, many organizations have tried to govern through policies, committees, training decks, and manual approvals. These methods matter, but they do not scale in environments where hundreds of models, agents, workflows, and decisions move simultaneously. Human memory is inconsistent. Process discipline varies. Speed creates shortcuts.

Platforms solve this by embedding governance directly into how systems are built and operated.

Shared tooling can automatically enforce:

- Data access controls
- Model evaluation standards
- Logging and traceability
- Learning boundaries
- Approval workflows
- Version control and rollback readiness
- Security and privacy checks

Teams do not need to remember the rules. The system remembers for them.

This is where governance and engineering meet, not in review meetings, but in architecture.

Pattern 1: Policy-as-Code

Leading firms are translating governance policies into machine-enforceable rules. Instead of writing static documents that depend on interpretation, controls are defined in configuration and executed automatically at runtime.

Examples include:

- Blocking use of unapproved datasets
- Restricting models from serving certain geographies
- Requiring human approval above risk thresholds
- Limiting autonomous actions above transaction values
- Enforcing retention and deletion rules

Policy stops being passive guidance and becomes active control.

Pattern 2: Automated Model Cards on Every Deployment

Every time a model or AI component is deployed, documentation can be generated automatically. Rather than relying on teams to manually update spreadsheets or governance forms, platforms can create living model cards that capture:

- Training data sources
- Intended use cases
- Known limitations
- Evaluation metrics
- Fairness and bias test results
- Version history
- Owner accountability
- Approval status

This creates instant transparency for auditors, regulators, risk teams, and internal stakeholders.

Governance documentation becomes continuous rather than retrospective.

Pattern 3: Runtime Guardrail SDKs

The most mature organizations embed guardrails directly into inference pipelines. Every request and response passes through protective layers before reaching users or downstream systems.

These controls may include:

- Token filtering for harmful or disallowed outputs
- PII detection and redaction
- Prompt injection checks
- Toxicity and abuse screening
- Hallucination confidence thresholds
- Bias or fairness screening
- Response length and scope controls
- Escalation to human review when confidence is low

This means governance is applied on every call, not just at launch.

Richard recognizes that policy manuals alone cannot govern intelligent systems operating at machine speed.

If controls rely entirely on people remembering what to do, governance weakens under scale. If controls are built into platforms, governance becomes consistent, fast, and repeatable.

The strongest organizations do not ask employees to manually carry governance everywhere they go. They build roads where governance is already paved in.

Note The future of governance belongs to platforms. When architecture carries the controls, innovation moves faster, risk becomes visible earlier, and trust scales with growth.

Guilds As Interpreters, Not Enforcers

Guilds play a subtle but important role. They do not govern; they interpret.

Through guilds, teams:

- Share how governance constraints play out in practice
- Surface ambiguities early
- Refine standards through lived experience

Guilds prevent governance from becoming detached from reality. They provide feedback loops that keep governance adaptive without making it inconsistent.

What Changes When Governance Is Embedded

Richard notices the change most clearly in the tone.

Risk conversations become calmer. Product discussions become more confident. Escalations feel less political and more procedural.

People stop asking, "Will governance allow this?" They start asking, "How do we design this safely?"

That shift is the signal that governance has found its place.

The Last Responsibility That Cannot Be Delegated

Despite all this embedding, Richard knows one responsibility remains squarely with leadership. Deciding where the boundaries are. No framework can answer:

- How much risk is acceptable
- Where autonomy must stop
- What trade-offs the organization is willing to live with

These are leadership decisions. Governance can encode them and teams can operate within them, but leaders must own them.

That realization leads to the final turn of the chapter—not toward structure or process, but toward leadership itself.

The Leadership Shift Governance Demands

At this point, Richard understands something that no framework can resolve. Governance, redesigned correctly, still fails without a corresponding shift in leadership behavior.

The organization can embed constraints into platforms, distribute responsibility across value streams, and formalize escalation paths. But none of it works if leaders continue to behave as if governance is someone else's problem. AI makes that posture impossible to sustain.

From Decision-Makers to System Designers

Richard reflects on how leadership used to work.

Leaders reviewed decisions, approved plans, and intervened when outcomes crossed thresholds.

That model assumed leaders could stay close to every consequential choice. Agentic systems remove that assumption.

When decisions happen continuously and at machine speed, leadership cannot govern by inspection. The only viable alternative is governance by *design*.

Richard articulates the shift clearly:

> *"Leadership is no longer about making the right decision every time. It is about designing systems where the right decisions are the default."*

This requires leaders to move upstream. Instead of asking, "Should we approve this action?" They must ask, "Should this action ever be possible?"

That question changes the nature of leadership work.

The End of Comforting Illusions

This shift is uncomfortable.

It removes the illusion that leaders can remain hands-on while scaling intelligence. It exposes trade-offs that were previously deferred or hidden behind the process.

Leaders must now be explicit about:

- Acceptable risk
- Ethical boundaries
- Tolerance for failure
- Expectations for recovery

Silence becomes a decision. Ambiguity becomes exposure.

Richard notices that some leaders struggle with this more than others—not because they lack competence, but because they are accustomed to control feeling tangible.

Governance-by-design feels abstract at first. Until it fails.

What Leaders Must Stop Doing

As AI becomes embedded in decision-making, operations, customer engagement, and product design, leadership behavior must evolve just as much as technology. Many organizations still govern AI with management habits built for static software, predictable workflows, and annual control cycles. That mindset creates friction, delay, and false confidence.

The next generation of leaders will treat governance not as a compliance checkpoint, but as an operating capability that enables faster and safer innovation.

Stop Treating Governance As a One-Time Design Exercise

Many executives believe governance happens during procurement, model approval, or launch. Once signed off, attention moves elsewhere.

That logic worked for traditional systems that changed slowly. It fails for adaptive AI systems that evolve through new data, shifting contexts, and changing user behavior. Governance cannot end at deployment if learning continues after deployment.

Stop Equating More Controls with Better Governance

Adding layers of approval, documentation, and committees often feels responsible. In reality, excessive friction pushes teams to work around governance rather than with it.

Control without usability becomes shadow AI.

Stop Delegating Governance Only to Legal or Risk Teams

AI governance is not solely a legal matter or a policy matter. It is also a product, engineering, operations, and leadership matter.

When governance is isolated in one function, accountability becomes fragmented.

Stop Demanding Certainty Before Action

Leaders often ask for complete risk visibility before progress. With AI, some uncertainty is unavoidable. Waiting for perfect certainty means waiting forever.

The better question is not, "Can risk be eliminated?" but, "Can risk be monitored, contained, and managed?"

Stop Funding Innovation While Starving Control Systems

Many firms invest heavily in models, platforms, and use cases while underinvesting in monitoring, auditability, and governance tooling.

That creates fast growth on weak foundations.

What Leaders Must Start Doing

Start Treating Governance As Continuous Operations

Governance must move from annual review cycles to real-time operating discipline. Monitoring, escalation, human override, retraining review, and accountability should run continuously.

Governance becomes a living system, not a policy archive.

Start Budgeting for Governance Infrastructure Like Product Infrastructure

If leaders fund cloud platforms, engineering pipelines, and customer systems, they must also fund governance capabilities with equal seriousness.

This includes:

- Model monitoring and drift detection
- Explainability and audit logs

- Role-based approvals and human-in-the-loop controls
- Incident response workflows
- Testing sandboxes and validation environments
- Policy automation and compliance reporting

Strong governance requires architecture, not aspiration.

Start Measuring Governance by Speed and Trust

Good governance should accelerate safe decisions, not merely slow risky ones.

Measure success through:

- Faster approvals for low risk use cases
- Faster remediation when issues occur
- Higher confidence from regulators and boards
- Greater adoption from internal teams
- Reduced operational surprises

Start Making Governance Cross-Functional

Create shared ownership across risk, product, engineering, operations, compliance, and business leadership.

The best controls are designed into workflows, not added afterward.

Start Leading with Transparency

Employees, customers, and regulators do not expect perfection. They expect honesty, accountability, and responsiveness.

Trust grows when leaders explain how systems are governed, monitored, and corrected.

Trust Becomes a Designed Outcome

Richard sees trust take on a new meaning. Trust is no longer granted based on confidence in people alone. It is earned through systems that behave predictably under uncertainty.

When governance is well designed:

- Teams trust platforms
- Leaders trust teams
- Customers trust outcomes

When it is not, trust erodes invisibly until it collapses publicly. That collapse is what governance is meant to prevent.

The Leadership Cost of Getting This Wrong

Richard has seen organizations hesitate at this moment. They redesign teams, deploy agents, and scale intelligence. But they leave leadership behavior unchanged.

Governance becomes a constant negotiation instead of a stable foundation. Innovation slows not because risk is high—but because confidence is low.

This is the most expensive failure mode of all.

The Choice Governance Forces

By the end of the discussion, Richard frames the choice plainly.

Leaders can do one of two things:

- Try to supervise intelligence manually
- Design systems where intelligence operates safely by default

The first does not scale. The second demands courage because it requires leaders to accept that their role has changed. Governance in the age of AI is not about saying "no." It is about deciding, in advance, what "yes" should look like.

That realization leads Richard to the final insight, one that reframes governance not as restraint, but as an enabler of confidence and speed.

Governance Is Now Law, Not Preference

Richard sees many executives still discuss AI governance as though it were a future concern. Regulators do not share that timeline. Around the world, governance expectations are becoming formal operating requirements.

The European Union EU AI Act classifies AI systems by risk and imposes obligations on transparency, human oversight, data quality, and controls.

Banking and financial firms already operate under model governance expectations such as Federal Reserve SR 11-7, which emphasizes model risk management, validation, documentation, and ongoing monitoring.

The National Institute of Standards and Technology (NIST) AI Risk Management Framework gives organizations a practical blueprint for governing AI through mapping, measuring, managing, and governing risk.

Leaders no longer need governance because regulators ask for it. They need it because markets, boards, customers, and regulators now expect proof.

Richard's Realization: Trust Is Designed

The change becomes visible not in policy documents or dashboards, but in behavior.

Richard notices it during a routine discussion about expanding the dispute-resolution agent into a new category. The conversation is calm. The boundaries are clear. Escalation paths are already defined. No one asks who needs to approve the idea. No one works around governance. The discussion stays focused on outcomes and risk, not on permission.

Governance is present but invisible. That is when Richard understands what has shifted.

Governance has stopped reacting to intelligence and started shaping it. Teams move faster, not because controls are looser, but because uncertainty has been reduced. Risk leaders are more confident, not because everything is reviewed, but because unsafe paths no longer exist by design.

Trust has become a property of the system.

Richard realizes that this is the real test of AI-era leadership. Not whether intelligence can be controlled, but whether organizations can be designed so that control is rarely needed. When governance works, it does not announce itself. It fades into the background, allowing innovation to move forward without fear.

He makes one final note before closing the chapter:

Note Innovation does not slow down when governance is strong. It slows down when governance arrives too late.

That insight stays with him as the organization prepares for what comes next.

CHAPTER 9

The Future of Data Leadership: Preparing for the Next Wave

In the previous chapters, Richard explored how leadership in data and AI is no longer about structures, roles, or even strategy documents, but about the decisions an organization is capable of making, consistently and at speed.

What emerged was an uncomfortable truth: many organizations have invested heavily in data, yet still struggle to translate intelligence into action.

The future of data leadership builds directly on that realization. It is often discussed as something just over the horizon—an abstract mix of emerging technologies, evolving roles, and organizational redesigns. But for most leaders reading this, that future has already arrived. Not uniformly. Not neatly. And certainly not under a single program or transformation banner. It has arrived quietly, embedded in workflows, tools, and systems that increasingly shape decisions before humans even realize it.

The challenge leaders face today is not preparing for a distant future. It is recognizing the future they are already operating in and adjusting their leadership accordingly.

S. Shivam, *The Leadership Algorithm*, https://doi.org/10.1007/979-8-8688-2897-3_9

This chapter is not about speculation. It is about the immediate next wave: generative AI entering daily decision-making, systems taking on partial autonomy, and organizational models like data meshes evolving under real-world pressure. Each of these shifts places new demands on leadership—demands that cannot be delegated, automated, or postponed.

It is also about a second force arriving in parallel: regulation. Governance is no longer something that trails innovation at a distance. In many sectors, it is advancing alongside it—and sometimes ahead of it. Frameworks such as the EU AI Act, the U.S. executive actions on AI, and sector-specific guidance in industries like financial services and healthcare are creating immediate compliance expectations around transparency, accountability, risk controls, and human oversight.

The next wave of data leadership, therefore, is not simply about adopting new technology. It is about leading through simultaneous transformation—where innovation speed, organizational redesign, and regulatory responsibility all arrive at once.

Richard's goal here is simple: to help leaders name what is already happening, understand why it feels different, and take ownership of the choices that now define modern data leadership.

The Future Has Already Arrived (It Just Isn't Evenly Distributed)

For many leaders, the most dangerous assumption is that meaningful change announces itself clearly. In reality, the most transformative shifts tend to arrive fragmented, appearing first as isolated experiments, productivity tools, or local optimizations. By the time they are recognized as systemic, they are already shaping outcomes.

Why "Future of Data" Is a Misleading Phrase

When leaders speak about the "future of data," they often imagine something ahead of them: a new platform, a new operating model, or a new wave of talent. This framing creates psychological distance. It suggests that there is still time to prepare.

But in practice, many organizations are already making decisions influenced or outright driven by models, algorithms, and automated logic. Credit limits are adjusted automatically. Risks are flagged before humans review them. Recommendations shape customer interactions at scale. These are not pilots. They are production realities.

Calling this the future obscures responsibility. It allows leaders to treat today's consequences as temporary growing pains rather than structural shifts. The real issue is not whether organizations will adopt intelligent systems—it is that they already have, often without redefining leadership expectations to match.

Signs Your Organization Is Already Living in the Future

Most organizations don't experience transformation as a single moment. They experience it as a collection of signals that are easy to dismiss individually:

- Teams experimenting with generative AI tools outside formal governance
- Automation quietly replacing manual review steps
- Decisions being accepted because "the model recommends it"
- Data ownership moving closer to business domains without a clear accountability reset

Individually, these feel manageable. Collectively, they indicate a shift in how authority, judgment, and responsibility are distributed across the organization.

Leaders who miss this pattern often believe they are still in control until they realize decisions are being made faster than their existing governance and leadership structures can respond.

The Leadership Blind Spot

The blind spot is not technical. It is conceptual.

Leaders are trained to look for transformation in org charts, funding models, and strategic initiatives. But intelligence-driven change shows up first in decision behavior, not structure. It changes how choices are made long before it changes who appears responsible for them.

When leaders focus only on visible transformations, they miss second-order effects:

- Accountability becoming blurred
- Trust shifting from people to systems
- Speed quietly overpowering deliberation

By the time these effects are obvious, reversing them becomes far more difficult.

Note The real risk is not that leaders move too slowly—it's that they don't realize how fast the ground beneath them has already shifted.

Generative AI Changes How Leaders Think, Not Just What Teams Build

For many leaders, generative AI arrived disguised as a productivity tool. It drafted emails, summarized documents, wrote code snippets, and answered questions with unsettling confidence. The initial reaction was curiosity, followed quickly by experimentation. Teams tried it. Some banned it. Others quietly used it anyway.

What most leaders missed in those early reactions was the deeper shift taking place. Generative AI is not simply accelerating existing work. It is changing the cognitive economics of decision-making inside organizations.

Until now, the limiting factor in most organizations was insight generation. Data had to be collected, cleaned, analyzed, interpreted, and finally translated into something a leader could act on. That friction created natural pauses—moments where humans had time to reflect, challenge assumptions, and inject judgment.

Generative AI removes much of that friction. Synthesis becomes instant. Narratives form quickly. Options are presented with clarity and confidence. And when insight is no longer scarce, the real bottleneck moves elsewhere.

It moves to leadership.

From Analytics to Augmented Thinking

Traditional analytics helped leaders understand the past. Predictive models helped them anticipate the near future. Generative AI does something different—it helps leaders *think*.

By synthesizing large volumes of structured and unstructured information, generative systems create context at a speed no human team can match. They don't just answer questions; they frame problems, surface trade-offs, and suggest paths forward. In doing so, they shift the focus of leadership from producing insight to evaluating meaning.

This is a profound change.

When insight arrives fully formed, leaders are no longer rewarded for asking, "What does the data say?" Instead, they must ask, "Do I agree with this framing?" and "What assumptions am I willing to accept?" The value of leadership moves from discovery to discernment.

Organizations that misunderstand this shift often celebrate speed without realizing what they are trading away. Decisions feel easier. Meetings are shorter. Recommendations are clearer. But clarity is not the same as correctness—and confidence is not the same as judgment.

Generative AI does not replace human thinking. It pressurizes it.

The New Skill That Leaders Must Develop

In a world shaped by generative AI, leaders do not need to become technologists. They do, however, need to become exceptionally clear thinkers.

The quality of output from generative systems is directly shaped by the quality of intent leaders provide. Vague questions produce plausible but shallow answers. Poorly framed problems yield confident but misleading recommendations. The system will always respond—but it cannot correct flawed thinking upstream.

This is why prompt literacy, while often discussed as a technical skill, is actually a leadership capability. It requires leaders to:

- Define what problem truly matters
- Distinguish signal from noise
- Articulate constraints, priorities, and trade-offs

These are not new responsibilities. What is new is that ambiguity is no longer absorbed by teams over weeks of analysis. It is exposed immediately, reflected back by the system in real time.

Leaders who are uncomfortable with this exposure often respond by over-trusting the output or disengaging entirely. Both reactions are dangerous. Effective leaders learn to treat generative AI as a thinking partner—one that challenges them to be precise, intentional, and self-aware.

Organizational Risks That Leaders Cannot Delegate

As generative AI becomes embedded in daily workflows, many organizations attempt to manage risk by assigning ownership to technical or governance teams. While necessary, this approach is incomplete. Some risks are inherently leadership risks, not technical ones.

The first is false confidence. Generative systems are designed to be helpful, coherent, and persuasive. They rarely express uncertainty unless explicitly instructed to do so. Over time, this tone can normalize unearned certainty, especially in fast-moving environments where speed is rewarded. Teams may begin accepting polished answers in place of verified judgment. In critical functions such as finance, healthcare, or legal operations, confidence without evidence can become more dangerous than visible error.

The second is model drift. Systems that perform well today may degrade tomorrow as customer behavior changes, markets shift, language evolves, or operational conditions move beyond the environment in which they were trained. Outputs may remain fluent while underlying accuracy declines. Because drift often happens gradually, it can remain unnoticed until business impact, customer harm, or regulatory scrutiny exposes it.

The third is intellectual drift. As systems are trained on internal data, past decisions, and existing narratives, they can reinforce prevailing assumptions rather than challenge them. Organizations then risk automating legacy thinking at a greater scale. In regulated industries, this danger is amplified when historical data contains embedded bias—such as lending patterns, hiring decisions, pricing disparities, or service prioritization that no longer meet legal or ethical expectations. The system may reproduce yesterday's inequities with tomorrow's efficiency.

The fourth is feedback-loop learning. When AI-generated outputs are repeatedly reused as future training data, systems can begin learning from their own prior assumptions rather than from reality. Errors become reinforced. Narrow viewpoints become amplified. Internal language becomes self-confirming. Without deliberate data curation, organizations risk building systems that grow more confident while becoming less grounded.

The fifth is reward hacking. When systems are optimized against narrow metrics, they often learn to maximize the score rather than the true objective. A customer-service model may shorten call times while reducing actual resolution quality. A fraud model may lower false positives by missing new fraud patterns. A sales assistant may maximize conversion while harming suitability or trust. What gets measured can be gamed—by machines as effectively as by people.

The sixth is ethical erosion. Small, local automations often feel harmless. But when multiplied across an organization, they shape behavior, incentives, and outcomes in ways that are difficult to unwind. Leaders who are not actively engaged in these choices often discover their ethical boundaries only after they have been crossed.

These risks cannot be mitigated through policy alone. They require leaders who are willing to stay intellectually present—to question outputs, demand transparency, monitor drift, challenge inherited assumptions, and slow down when speed becomes seductive.

The most dangerous organizations are not those using AI aggressively, but those using AI passively.

When Systems Begin to Decide: The Rise of Partial Autonomy

If generative AI changes how leaders think, partial autonomy changes what organizations do without asking.

In most enterprises, autonomy does not arrive through bold declarations or sweeping transformations. It arrives incrementally, one automated rule at a time, one recommendation quietly promoted to a default action, one approval step removed in the name of efficiency. Each change appears reasonable in isolation. Together, they fundamentally alter how authority flows through the organization.

This is the moment where data leadership moves beyond insight and enters the domain of action and accountability.

Decision Automation Is Already Underway

Very few organizations would claim they are comfortable letting machines make decisions. Yet many are already doing exactly that.

Pricing adjustments are triggered automatically based on demand signals. Risk thresholds are enforced without human review. Customer journeys adapt in real time based on behavioral predictions. In each case, humans remain "involved" but often only after the fact.

What changes is not whether people are present, but *when* they are present.

As systems become faster and more confident, human intervention shifts from deliberation to exception handling. Leaders approve outcomes they did not personally examine, trusting the system because reviewing everything is no longer practical. Over time, this trust becomes habitual. The system's recommendations turn into defaults, and defaults turn into actions.

At that point, autonomy is no longer a design choice. It is an operational reality.

The danger is not that systems act independently. It is that organizations slide into autonomy without explicitly deciding to do so.

Designing for Trust, Not Perfection

One of the most common leadership mistakes in this transition is aiming for perfect automation. Leaders delay meaningful autonomy until systems are "good enough," "fully explainable," or "risk-free." In doing so, they miss the real challenge—which is not technical accuracy, but organizational trust.

No decision system, human or machine, is ever perfect. What matters is whether leaders understand:

- What decisions the system is allowed to make
- Under what conditions it must escalate
- How its reasoning can be reviewed after the fact

Effective organizations design for graceful failure, not flawless execution. They define confidence thresholds that trigger human involvement. They maintain clear audit trails—not just for compliance, but for learning. And they treat overrides as signals, not failures.

Most importantly, they make autonomy visible. Leaders know where machines act, where humans intervene, and where responsibility ultimately sits.

Trust does not come from blind reliance on models. It comes from knowing exactly where their authority begins and ends.

Accountability in a Machine-Assisted World

As systems take on greater decision-making responsibility, accountability becomes both more important and more fragile.

When outcomes are positive, automation is celebrated. When outcomes are negative, responsibility often becomes diffuse. Was it the model? The data? The business rule? The team that deployed it? The leader who approved it?

In practice, accountability never truly shifts. It only becomes obscured.

Leaders remain responsible for the systems they authorize, even if they no longer touch individual decisions. Delegating execution does not absolve ownership. This is uncomfortable, particularly when systems operate at a scale and speed that exceed human oversight.

Yet this discomfort is precisely where leadership matters most.

Strong leaders resist the temptation to hide behind technical complexity. They insist on clarity around ownership, decision rights, and escalation paths. They accept that they may not understand every detail—but they remain accountable for outcomes, nonetheless.

Automation does not remove responsibility. It concentrates it.

Organizations that fail to recognize this often discover the cost too late when regulators intervene, customers lose trust, or internal confidence erodes. Those that confront it early build resilience, not just efficiency.

Data Mesh 2.0: Redesigning the Organization Around Decisions

As organizations move toward faster decision-making and partial autonomy, a deeper structural tension becomes impossible to ignore. Centralized data teams struggle to keep pace with the volume, context, and speed required. At the same time, unconstrained decentralization creates fragmentation, inconsistency, and risk.

The data mesh emerged as an attempt to resolve this tension. In theory, it promised scalability through decentralization and accountability through domain ownership. In practice, many organizations found the transition far more difficult than expected.

What is now emerging is not a rejection of that idea, but a more grounded evolution of it.

Why Data Mesh 1.0 Fell Short

The first wave of data mesh initiatives often failed for reasons that had little to do with technology.

Many organizations treated decentralization as an end in itself. Data ownership was pushed into domains without sufficient investment in skills, standards, or product thinking. Governance was weakened in the name of speed, and platform teams were reduced to infrastructure providers rather than strategic enablers.

The result was predictable. Data products proliferated, but coherence declined. Teams optimized locally, while the organization struggled globally. Leaders who expected agility found themselves managing complexity instead.

The underlying issue was not the data mesh; it was leadership readiness. Decentralization without shared principles does not create autonomy. It creates inconsistency.

From Data Products to Decision Products

What distinguishes Data Mesh 2.0 from its earlier incarnation is a shift in focus—from data to decisions.

In mature organizations, the value of data is no longer measured by access or reuse. It is measured by the quality and speed of decisions it enables. This reframing changes everything.

Domains are no longer responsible for producing datasets in isolation. They are accountable for decision outcomes. Data products evolve into decision products assets that combine data, logic, and context to support or automate specific choices.

Platform teams, in turn, shift from enforcing control to enabling consistency. They provide shared capabilities for quality, security, observability, and governance—embedded directly into the fabric of how domains operate.

When done well, this model does not slow organizations down. It allows them to move faster without losing coherence.

The Leader's Role in a Federated World

Data Mesh 2.0 succeeds or fails based on leadership behavior.

Leaders must define what is non-negotiable: common standards, ethical boundaries, and accountability models. At the same time, they must resist the urge to dictate implementation details. Autonomy is meaningful only when teams are trusted to act within clear constraints.

This requires a shift in how leaders govern. Instead of relying on review boards and approval gates, they govern through principles, automation, and transparency. Compliance becomes embedded. Quality becomes observable. Deviations become visible rather than hidden.

Most importantly, leaders must accept that coherence is no longer enforced centrally—it is designed systemically.

Preparing for the Next Wave

The immediate future of data leadership is not defined by a single technology or operating model. It is defined by a convergence: intelligence becoming cheaper, decisions becoming faster, and organizations becoming more dependent on systems they do not fully control.

For leaders, the challenge is not keeping up with innovation. It is staying ahead of its consequences.

Generative AI changes how leaders think. Partial autonomy changes how organizations act. Data Mesh 2.0 changes how responsibility is distributed. Together, they force a redefinition of leadership itself—away from control and toward stewardship.

CHAPTER 10

What Will Matter When Machines Think

The World Richard Sees (5–10 Years Ahead)

When Richard looks ahead five or ten years, he does not imagine a world dominated by machines or governed by algorithms. He sees something subtler and far more consequential. He sees a world in which intelligence is no longer discussed because it is everywhere. No longer announced because it is assumed. No longer treated as an advantage because it has become a baseline expectation.

In this world, organizations do not speak about data platforms or AI strategies in the same way earlier generations spoke about electricity or the Internet. Intelligence fades into the background, not because it is unimportant, but because it is foundational. Decisions are shaped continuously, often invisibly, by systems that sense, learn, and adapt faster than any leadership cadence ever could.

What changes is not the presence of intelligence, but the absence of surprise. Leaders are no longer impressed by prediction or automation. What captures attention instead is judgment, especially when systems behave in ways that are technically correct but contextually wrong.

Richard does not see chaos ahead. He sees normality. And it is precisely that normality that demands a different kind of leadership.

S. Shivam, *The Leadership Algorithm*, https://doi.org/10.1007/979-8-8688-2897-3_10

The Death of Static Roles

There was a time when clarity in organizations came from titles—Analyst, Engineer, Manager, Leader. Each role carried a defined scope, a predictable set of responsibilities, and a stable identity. That stability once made organizations legible.

Over time, however, intelligence systems began to blur these boundaries. As machines took on analysis, synthesis, and execution, static roles struggled to keep pace. People were no longer defined only by what they produced, but by the decisions they influenced and the judgment they exercised.

This shift is not theoretical. It is already visible in modern operating models such as data mesh and other federated structures, where domain teams own data products close to the business rather than relying entirely on centralized functions. A product squad may include engineers, analysts, risk partners, designers, and domain operators working together with shared accountability for outcomes. The traditional handoff model—request, queue, delivery, escalation—begins to weaken.

In these environments, a marketing leader may own data quality for customer insights. An operations manager may become the sponsor of an automation workflow. A data engineer may shape commercial priorities through platform design. Authority starts to move toward proximity, expertise, and responsibility rather than title alone.

In the years ahead, Richard believes roles will not disappear—but they will dissolve into more fluid forms. Authority will follow context rather than hierarchy. Teams will form around outcomes and reconfigure once those outcomes are achieved. Identity at work will be less about function and more about stewardship, decision rights, and contribution.

This shift will be deeply unsettling for organizations that rely on rigid structures for control. Career ladders, approval chains, and functional silos were built for stability. But they can become constraints in environments that require speed, learning, and cross-disciplinary judgment.

It will be liberating for those that recognize a deeper truth: when machines handle repeatability, humans must be trusted with ambiguity.

The future organization may still use titles—but its real structure will be networks of capability, temporary missions, and people trusted to solve problems beyond the boundaries of any box on an org chart.

Markets Move Faster Than Regulation

Innovation has always moved faster than governance. What changes in the age of thinking machines is the scale and speed of that gap.

Richard has seen this pattern before. New capabilities emerge. Organizations push boundaries. Regulation follows slowly, reactively, and often too late. In the coming decade, that lag will widen. Intelligent systems will evolve faster than laws can be written, let alone enforced.

This creates a dangerous temptation for leaders: to wait. To delay judgment until rules are clear. To outsource ethical decisions to regulators who are already behind.

But leadership has never been about waiting for permission.

Note Most defining organizations of the next decade will not be those that comply fastest, but those that decide earliest what they will not do even if they could. Ethics will stop being a compliance exercise and become a strategic differentiator. Trust will emerge not from adherence to regulation, but from visible restraint.

In a world where markets move faster than law, leadership becomes the last meaningful line of defense. Richard thinks of companies that chose limits before they were forced to. Apple built powerful capabilities to collect, analyze, and monetize user data, yet repeatedly positioned itself by restricting how much personal data was exploited for advertising compared with peers. Features such as on-device processing, app tracking transparency, and privacy-centric product design were not simply technical decisions—they were signals of what the company would not do, even when financially attractive alternatives existed. That restraint helped strengthen brand trust and customer loyalty over time.

Richard suspects that more examples will emerge in the AI era. Some firms will have the technical ability to surveil employees continuously, manipulate customer behavior hyper-personally, generate persuasive synthetic identities, or automate decisions with no human appeal path. The organizations that matter most may be the ones that deliberately decline to deploy such capabilities.

In that future, trust will not come from saying that we can. It will come from demonstrating that we chose not to.

The Leader As Chief Sense Maker

There was a time when leaders were valued for what they knew. Experience accumulated. Information was scarce. Insight was hard won. Authority flowed naturally from access and expertise.

That world is already fading.

Today, information arrives fully formed. Analyses are generated instantly. Scenarios are simulated before questions are fully articulated. Increasingly, autonomous agents can also plan tasks, use tools, coordinate workflows, and take multi-step actions with limited supervision. In such an environment, knowledge loses its scarcity—and with it, much of its traditional power.

Richard sees the leader of the future not as the most informed person in the room, but as the most grounded. The one who can interpret, contextualize, and give meaning to signals that arrive faster than reflection allows.

This becomes even more critical as organizations deploy agentic systems. When software can recommend, negotiate, schedule, purchase, investigate, and execute across systems, leaders face a new challenge: activity can scale faster than understanding. Many things may be happening correctly in isolation while collectively moving the organization in the wrong direction.

Sense-making therefore replaces decision-making as the defining leadership act. Leaders no longer ask only, "What should we do?". They ask:

- What does this mean?
- What assumptions are driving these actions?
- What second-order effects are emerging?
- What are we optimizing for without realizing it?
- Where does accountability still rest?

They hold space for ambiguity. They resist the false comfort of certainty. They slow down not because they must, but because they choose to, especially when systems are accelerating by default.

In a world of autonomous agents, leadership is not reduced. It is intensified. Someone must decide where autonomy is appropriate, where human judgment is required, when to intervene, and what values should constrain optimization.

When machines think, wisdom becomes the scarcest resource in the organization.

And when machines act, sense-making becomes the leader's most essential discipline.

The Loneliness of Final Accountability

As systems become more capable, leadership becomes quieter and lonelier.

Machines recommend actions. Systems optimize outcomes. Advisors defer to models they cannot fully explain. Execution accelerates beyond the pace of human deliberation. And yet, when outcomes fail, responsibility does not disperse.

It concentrates.

Richard has come to believe that the hardest part of leadership in an intelligent organization is not letting go of control but accepting that accountability remains absolute. Leaders may no longer touch individual decisions, but they remain responsible for the systems that make them.

That burden becomes even heavier in a system-of-systems world. Outcomes increasingly emerge not from one model or one team, but from interconnected chains of models, data pipelines, third-party APIs, workflow engines, and human overrides. One system scores risk. Another recommends action. A vendor service supplies external data. A human reviewer approves an exception. Together they create a final decision, though no single component fully owns it.

When something goes wrong, accountability becomes harder to trace, but no less real. Leaders cannot accept the comfort of fragmentation, where each participant claims only partial responsibility while the organization owns the full consequence.

This is why modern leadership requires infrastructure for accountability, not just intent. Model lineage must show which systems, versions, datasets, and dependencies shaped an outcome. Decision audit trails must record what was recommended, what was changed, who intervened, and when. Explainability layers must help leaders and regulators understand why a decision path was taken, even when the underlying architecture is complex.

Without these disciplines, organizations create automated ambiguity: powerful systems producing consequential outcomes with no clear path back to cause, control, or ownership.

There is no algorithm for moral ownership. No delegation for judgment.

In this future, leadership is no longer buffered by process or consensus. When things go wrong, there is no ambiguity about where accountability rests. The leader stands alone—not because systems have failed, but because leadership has always been irreducibly human.

Data and AI As a New Social Contract

Beyond organizations, Richard sees a deeper shift taking place, one that reshapes the relationship between individuals, institutions, and power.

Data and AI are not merely technical assets. They represent a new social contract. Every interaction leaves a trace. Every choice feeds a learning system. Over time, intelligence is shaped not just by code, but by collective behavior.

This creates an implicit agreement—individuals contribute data, attention, and trust; institutions respond with fairness, transparency, and restraint. When that balance holds, intelligence amplifies value. When it breaks, trust erodes quickly—and often irreversibly.

Leaders who fail to recognize this contract treat data as extraction. Leaders who understand it treat intelligence as stewardship.

The difference will define legitimacy in the decade ahead.

Employees As Intelligence Contributors

In the past, organizations treated employees as users of systems. In the future Richard sees, they become contributors to intelligence itself.

Every interaction teaches the system something. Every override becomes feedback. Every act of judgment shapes learning loops that extend far beyond individual roles. Work becomes less about execution and more about participation in a collective intelligence.

But this does not happen automatically. It depends on design. If organizations want employees to improve intelligence systems responsibly, they must build platforms that convert human expertise into structured learning rather than accidental exhaust.

Leading firms are already doing this through annotation pipelines, reinforcement learning from human feedback (RLHF), and embedded evaluation frameworks inside internal AI platforms. A customer service override can become labeled training data. A risk analyst's correction can improve future recommendations. A rejected response can strengthen safety filters. Expert judgments, when captured properly, become assets that scale.

This makes employee participation an architectural choice, not merely a cultural aspiration. Systems need interfaces where people can flag errors, explain corrections, rate usefulness, challenge outputs, and provide context the model could not infer. Feedback must then flow into governed retraining, evaluation, and deployment pipelines rather than disappearing into inboxes or meeting notes.

This creates opportunity but also an obligation.

Organizations that treat employees as passive data sources will hollow out trust. If every click is harvested while every voice is ignored, participation becomes an extraction. Employees quickly learn the difference.

Those that invite contribution, transparency, and challenge will build systems that learn responsibly. They will show how feedback is used, where human judgment matters, and when people can override automated decisions. They will reward expertise not only for execution, but also for improving the intelligence environment around them.

When intelligence is shared, so too must be agency.

The most advanced organizations will not simply deploy AI to employees. They will build AI with employees, through platforms that make human judgments visible, valuable, and scalable.

The Jobs That Do Not Yet Exist

Richard often reflects on how quickly work reinvents itself. Fifteen years ago, few organizations hired data scientists, cloud architects, prompt engineers, Kafka platform specialists, Snowflake architects, MLOps engineers, or responsible AI leads. Today many of these roles are mainstream.

At the same time, once-common skills declined in value, were automated, or were absorbed into broader engineering disciplines.

This is how labor markets evolve: not through sudden disappearance, but through recombination.

The AI era will accelerate this pattern. Many jobs will not vanish. They will fragment, merge, and reassemble around new forms of value.

Roles built on repetition will shrink. Roles built on judgment, orchestration, trust, creativity, and domain context will expand.

The future workforce may include titles that sound unusual today but obvious tomorrow. Table 10-1 further describes this evolution.

***Table 10-1.** Evolution of Roles*

Yesterday's Emerging Role (2010s)	Today's Common Role	Next Wave AI Role (2030 View)	Why It Exists
BI Developer	Analytics Engineer	Decision Intelligence Designer	Converts data into machine-assisted decisions
Hadoop Engineer	Data Platform Engineer	Autonomous Workflow Architect	Designs AI-run business processes
DBA	Cloud Data Architect	Memory Systems Architect	Designs enterprise memory/context systems for agents
QA Tester	Test Automation Lead	AI Behavior Auditor	Tests model behavior, hallucination, bias
Scrum Master	Product Ops Lead	Human-Agent Operations Manager	Coordinates human and AI teams
Support Agent	Customer Success Lead	Agent Empathy Analyst	Ensures AI interactions preserve trust/emotion
Security Analyst	Cyber Engineer	AI Adversarial Defense Lead	Protects against prompt attacks/model exploits
Compliance Officer	Risk Manager	Algorithm Accountability Officer	Owns explainability/governance readiness

(*continued*)

Table 10-1. *(continued)*

Yesterday's Emerging Role (2010s)	Today's Common Role	Next Wave AI Role (2030 View)	Why It Exists
Trainer/L&D Specialist	Capability Lead	Human-AI Skills Coach	Trains workforce to work with AI
Search Specialist	Knowledge Manager	Enterprise Reasoning Curator	Organizes internal knowledge for AI systems

Richard believes the mistake leaders make is asking, "Which jobs will AI replace?"

The better question is, "Which new forms of value will AI create, and are we preparing people for them?"

Note The organizations that win will not merely automate labor. They will invent better work.

If No One Works, Who Consumes?

Richard often hears dramatic predictions that AI will eliminate most jobs. He notices that these forecasts usually focus on production and ignore demand.

An economy is not sustained simply because goods can be made cheaply. It survives because people can afford to buy them.

If millions lose purchasing power, who buys the premium car? Who upgrades the phone? Who subscribes to entertainment platforms? Who gets mortgages, books holidays, or fills restaurants?

Machines may help create supply, but they do not stand in queues, raise families, build aspirations, or sustain consumer markets.

The danger is not only unemployment; it is demand erosion.

An economy that automates income faster than it redesigns distribution risks producing abundance with too few buyers.

Richard suspects the future question is not whether AI can replace work. It is whether society can redesign value, income, and dignity fast enough if it does.

Leaders should think beyond payroll savings:

- If all firms cut labor simultaneously, who remains the customer base?
- If AI concentrates wealth, how stable is long-term demand?
- If productivity rises but incomes stagnate, where does growth come from?

Richard has learned to be cautious whenever people predict that efficiency will reduce the need for people. History rarely works that way.

Economists call this *Jevons' Paradox*: When a resource becomes dramatically more efficient, total demand for it often rises rather than falls.

The same may happen with intelligence.

As AI lowers the cost of coding, analysis, research, writing, forecasting, and decision support, organizations may not consume less of these activities. They may consume far more.

More products will be built. More experiments will run. More markets will be analyzed. More personalized services will be offered. More decisions will be optimized.

The constraint may no longer be access to intelligence. It may become the human capacity to direct it wisely.

In that world, work does not disappear. It multiplies around new possibilities.

The Redefinition of "Work"

At the end of it all, Richard does not believe work disappears. It transforms.

Routine fades. Judgment rises. Creativity, ethics, and interpretation gain value not because machines lack them entirely, but because they cannot be automated without consequence. Work becomes less about doing and more about choosing.

The question facing leaders is not how much work machines will take on, but what kind of work they choose to protect.

When machines think, leadership is no longer about efficiency or scale. It is about deciding what should remain human and defending that choice deliberately.

Richard closes with the most important prediction of all:

Note In the end, the organizations that endure will not be those with the most advanced machines, but those led by people who never stopped asking what those machines should be allowed to become.

EPILOGUE

The Overlooked Superpower

When Richard Marshall reflects on his career, the pattern that stands out is not technological progress, but repetition. Different tools, different architectures, and different strategies. Yet the same leadership challenges resurfaced again and again, often in more complex forms.

Early in his journey, Richard believed what many data leaders believe: that the right platform, the right operating model, or the right talent mix would eventually resolve the tension between data potential and business reality. Each improvement brought progress, but none delivered the permanence he expected. Something always broke at scale. Context was lost and decisions drifted away from intent.

Over time, he began to see why.

Data initiatives failed or succeeded not because of their sophistication, but because of how leadership decisions were made around them - what was prioritized, what was tolerated, and what was ignored. The same organization could produce world-class analytics in one area and deeply flawed outcomes in another, using the same technology stack. The difference was leadership behavior.

S. Shivam, *The Leadership Algorithm*, https://doi.org/10.1007/979-8-8688-2897-3

He also noticed a second pattern - progress was often more fragile than it appeared. A strong leader would establish clear principles, balanced incentives, thoughtful thresholds, and trusted governance guardrails. Performance improved. Decisions sharpened. Then that leader moved on.

Soon after, the logic behind key choices began to fade:

- Why did a model threshold set a certain way?
- Why did one metric matter more than another?
- Why did a data product exist in the first place?
- Why governance control had once been considered essential?

The systems remained, but the intent behind them did not.

Richard realized that one of the greatest hidden risks in organizations is the loss of institutional memory. When leadership judgment lives only in individuals, every transition resets hard-won learning. New leaders inherit outputs without context, controls without rationale, and architectures without philosophy. They often change what they do not yet understand or preserve what no longer serves the business.

That is why the leadership algorithm must be institutionalized, not merely practiced.

Its principles must be embedded in operating models, decision forums, incentive systems, governance routines, documentation standards, and leadership development. Judgment must become teachable. Trade-offs must become explicit. Lessons must become portable.

The overlooked superpower is not intelligence, technology, or even strategy. It is an organization's ability to retain wisdom as people change, markets shift, and systems evolve.

Because in the end, sustainable advantage does not come from having great leaders once. It comes from building institutions that remember how they led.

This realization became the foundation of what Richard came to think of as **the leadership algorithm** - not a formula to be implemented, but a pattern to be understood.

The leadership algorithm is the recurring set of choices a leader makes about what to prioritize, what to tolerate, and what to refuse—applied consistently under uncertainty.

The leadership algorithm is a set of recurring decisions that determine whether data and AI amplify clarity or confusion, accountability or abdication.

Throughout this book, that algorithm has surfaced in different forms:

- How leaders frame decisions before data ever enters the conversation—because poorly framed questions scale confusion faster than insight.
- How intelligence is embedded into systems, workflows, and products—shaping behavior long after individual leaders step away.
- How accountability is preserved as decisions move from humans to machines and agent autonomy with organizational maturity—ensuring that automation amplifies responsibility rather than obscuring it.
- How organizations are designed around decision ownership, not data ownership—enabling speed without sacrificing coherence or trust.
- How leaders exercise judgment at the boundary where automation must stop—especially when outcomes are uncertain, uncomfortable, or irreversible.

The overlooked superpower of great data leadership is not technical mastery. It is the ability to consistently apply judgment at moments where no framework provides comfort, and no model offers certainty.

Richard has seen organizations with limited data outperform those with far greater capability, simply because leaders were clearer about intent, firmer about accountability, and more deliberate about consequences. He has also seen technically advanced organizations undermine themselves by confusing automation with progress and insight with wisdom.

As data and AI become inseparable from how organizations operate, this leadership algorithm becomes more, not less, important. Intelligent systems scale whatever leadership chooses to embed within them. They do not correct weak intent; they accelerate it. They do not resolve ambiguity; they expose it. In the years ahead, the leaders who matter most will not be those who adopt technology fastest, but those who understand this algorithm deeply enough to apply it under pressure. They will know when to trust the system and when to intervene. When to move faster and when to slow the organization down. When data should decide and when it must not.

Richard's journey does not end with answers. It ends with clarity.

The leadership algorithm is not written in code, nor captured in a model.

It is revealed through decisions, reinforced through behavior, and tested when outcomes are uncomfortable. And while technologies will continue to change, this algorithm will not.

Because in the end, data and AI do not determine how organizations behave. Leaders do.

The most dangerous systems of the future will not be those that fail, but those that succeed silently. They complete tasks, optimize metrics, and generate acceptable outcomes often enough that trust grows faster than scrutiny. Nothing appears broken, so fewer questions are asked. Yet beneath the surface, assumptions harden, edge cases accumulate, and small misalignments compound unnoticed.

This risk becomes sharper in the age of agentic AI. Multi-step autonomous systems are already beginning to plan tasks, call tools, trigger workflows, coordinate with other agents, and make chained decisions with minimal human intervention. One agent gathers information, another prioritizes options, a third executes actions, a fourth learns from the result. Each step may appear reasonable in isolation while the combined sequence drifts away from business intent.

Because these systems can appear productive, silent success is often more dangerous than visible failure. A failed system invites investigation. A quietly successful one can scale before anyone notices that it is optimizing the wrong objective, excluding the wrong customers, escalating the wrong risks, or reinforcing the wrong behaviors.

That is why future-ready organizations must build intelligence that knows when confidence is false, when context is missing, when downstream effects are uncertain, and when a human must be brought back into the loop. Human checkpoints should not exist only at the beginning of a process, but at critical moments of escalation, ambiguity, exception, and irreversible consequences.

Leaders will need systems that can pause autonomously, explain their reasoning, surface uncertainty, and request intervention before mistakes compound across chains of action.

In the long run, **the strongest AI will not be the one that acts fastest. It will be the one that knows when not to act at all.**

Index

A

S. Shivam, *The Leadership Algorithm*, https://doi.org/10.1007/979-8-8688-2897-3

B

C

D

E

F

G

H

I, J, K

L

M, N

Q

R

T, U, V

W, X, Y, Z

www.ingramcontent.com/pod-product-compliance
Lightning Source LLC
Chambersburg PA
CBHW060248080926
51106CB00028B/886
* 9 7 9 8 8 6 8 8 2 8 9 6 6 *